The Beginner's Guide to Worm Farms and Vermicomposting

Small-Scale Vermicomposting for Gardeners, Micro-Farms, Regenerative Homesteads, and Permaculture Projects

Madalina Podgorean

For information address
LP Media Inc. Publishing
1405 Kingsview Ln N,
Plymouth MN 55447
www.lpmedia.org

Publication Data

Madalina Podgorean
The Beginner's Guide to Worm Farms and Vermicomposting — First edition.

Summary: "Successfully starting and maintaining your first vermicomposting system" Provided by publisher.

ISBN: 978-1-961846-11-1

[1. The Beginner's Guide to Worm Farms and Vermicomposting — Non-Fiction] I. Title.

TABLE OF CONTENTS

1

INTRODUCTION TO VERMICOMPOSTING

What Exactly is Vermicomposting?

Increasingly, the world needs sustainable agricultural methods that are affordable, easy to manage in our own homes and gardens, and that make a real, measurable impact both in our own lives and in the world at large.

Vermicomposting happens to be one of the easiest and fastest methods to learn. The best part is that it can be done on a very small scale in a compact or urban environment, or on a very large scale, by entire farms dedicated to producing vermicompost for sale or large-scale agricultural use.

So, what is vermicomposting?

Vermicomposting is the process of using worms to decompose organic waste into nutrient-rich compost. It involves adding worms to a container of organic waste, which the worms then consume and convert into compost. Using the natural enzymes and bacteria in their digestive systems, the worms break down the waste matter you feed them and turn it into solid castings through a process called vermicomposting.

This solid or liquid substance can then be used to help your seeds germinate faster, your plants grow bigger and healthier, increase your overall harvest, and clean up the earth's environment. If it sounds too good to be true, it's because vermicomposting truly is a superhero helper for your plants, vegetables, and environment.

When done correctly and with a little research and education, vermicomposting is a clean and easy process that produces no offensive odors and very little mess. It thrives even when you involve the help of little hands, and your vermiculate friends will tolerate your two-week vacation much better than your household plants ever will.

Vermicomposting is equally beneficial for the seasoned gardener or the hopeful beginner and is an easy and fun way to connect to the earth and the bounties it offers.

FUN FACT

An Ancient Art

Worms have been recognized for their contributions to healthy soil for thousands of years. The Greek philosopher Aristotle called earthworms the "intestines of the earth" and was one of the first recorded people to recognize the value of these tiny invertebrates. Today vermicomposting is one of the most popular and efficient methods for recycling kitchen waste.

Benefits of Vermicomposting

There are numerous benefits to vermicomposting, such as raising the fertility of your soil, boosting the health of your plants and crops, reducing the amount of waste that ends up in landfills, and reducing greenhouse emissions. Vermicomposting has risen in popularity because it addresses each of these concerns and requires minimal time, money, and effort to start.

Audrey Wynkoop of The Worm Bucket suggests that vermicomposting can also act as a great way to educate others about the importance of reducing waste and taking care of the environment. You can use your worm farm as a way to teach others about the benefits of composting and the role that worms play in the process. Not to mention, it's a great way to help children connect to the earth and spend more time outdoors.

IMPROVING SOIL HEALTH AND FERTILITY

A highly nutritious amendment to your soil, vermicomposting done well can dramatically and quickly improve the health and fertility of your soil and everything growing in it. It transforms regular soil into supercharged soil, teeming with beneficial microbes that help your plants grow stronger and more resistant to pests and diseases.

It does all this while saving you money on fertilizers and soil amendments that you would otherwise have to purchase, and it leaves your soil clean of chemicals and products that many of us would rather not include on our dinner plates. The more you use naturally produced vermicompost organic matter, the faster your garden turns into a growing environment that doesn't need chemical pesticides at all. Your soil becomes hardier, more fertile, and less dependent on products you have to purchase.

We started our first vermicompost project on our nine-acre farm to increase the health of our soil. We didn't want to use chemical fertilizers or any other types of chemical products. We knew that everything we grew would end up on our dinner plates, so we were motivated to succeed in keeping our farm as clean and healthy as possible. Even so, we were faced with depleted, dry soil that had supported many years of abuse. When we moved onto our land, we tried to find worms in the soil. There were none to be seen. If they existed at one time, they had packed up and moved to more fertile lands long before we arrived.

Vermicompost was one of several methods we used to heal our soil, bit by bit.

Vermicompost Compared to Chemical Fertilizers

Vermicompost naturally contains high amounts of nitrogen, phosphorous, potassium, various beneficial microbes, and micro-nutrients. It also enhances "plant growth directly by production of plant growth-regulating hormones and enzymes and indirectly by controlling plant pathogens, nematodes and other pests." (Pathma, 2012). It is often compared to NPK (nitrogen, phosphorus, and potassium) chemical fertilizers because of its contents but without the various drawbacks of conventional chemical fertilizers. Since vermicompost is a wholly organic and natural process, it is also more readily absorbed and assimilated by your soil. "Vermi-compost is not easily flushed from the soil because of the worm mucus that it contains." (Gill & Dommalapati, 2020) This means it gets a chance to work for a longer period of time and doesn't have to be applied as frequently as chemical fertilizers.

VERMICOMPOST VERSUS CHEMICAL FERTILIZERS

	Vermicompost	Chemical Fertilizer
Yield and crop quality	Increases both yield and crop quality	Only increases yield and has no effect on plant health
Nutrient absorption and availability	More easily absorbed by the plants, as all nutrients are completely natural and organic	Contains only several ingredients compared to the many nutrients and microorganisms found in vermicompost Easily flushed away from soil, possibly necessitating multiple applications during the growing season
Nutritional profile	Contains many nutrients, such as plant growth hormones, micronutrients, and beneficial organisms	Only a few ingredients in each type of chemical fertilizer, usually nitrogen, phosphorus, and potassium
Financial cost	Completely free if you feed your worms kitchen scraps and waste from your animals Production of vermicompost can increase dramatically in a matter of months without any additional costs.	Starting around $30 for a small bag—and more for organic fertilizers.
Resistance to pests and disease	Reduces the need for chemical fertilizers both in the short and long term and leads to healthier and more pest-resistant plants	Does not protect against disease and pests Eventually harms the soil, requiring even more chemical interventions long term
Possible problems	Must maintain vermicompost containers at proper temperature and moisture level, so the health of your worms doesn't suffer	Can easily apply too much fertilizer, harming your plants Will still have to deal with pests and disease separately

REDUCING WASTE

Food waste is a problem almost everywhere in the world and even more so in North America, Europe, and other affluent parts. Some of this food comes from our homes and restaurants, some is lost through spoilage even before it reaches its intended destination, and some is lost through the process of agriculture in the form of ruined or unharvested crops. Altogether, it makes up a substantial portion of the waste that ends up in landfills.

In "Sustainable Management of Food Basics" (2023), The U.S. Environmental Protection Agency estimated that in 2018 in the United States, more food reached landfills and combustion facilities than any other single material in our everyday trash. That's substantial, considering how much total waste we produce every year.

However, it doesn't all have to end up in landfills. In our home, among several adults and two children, we produce at least a small bucket of kitchen waste every day. We grow and eat a lot of fruits and vegetables and drink a few coffees each day. That's in addition to the vast amount of yard waste that our nine-acre farm produces. Only so much banana foliage can be used as mulch. We still have plenty of organic material left over.

Our hardworking worms do a great job of reducing that waste. The list of things your worms will happily eat is extensive: fruits and vegetables and their peels, coffee grounds, eggshells, tea leaves, all sorts of yard waste, scrap paper and cardboard, and manure from different types of livestock. Plus, your worms will help you reduce all this waste with minimal intervention on your part. They do most of the work, and your garden and the earth reap the benefits.

> "One of the biggest benefits of vermicomposting is eliminating the amount of garbage that heads out to landfills. Worms will consume a lot of table scraps and cardboard, eliminating the need to throw those items in the trash and eventually get dumped into a landfill. That's a benefit both locally and for the world at large!"
>
> **Ken Mitzel**
> Hobby Worm Farm

REDUCING GREENHOUSE GAS EMISSIONS

On a larger scale, worms are doing their part in reducing the effects of climate change. Vermicomposting can help bring down greenhouse gas emissions by reducing the amount of methane produced in landfills. This is especially true for the fresh food scraps that end up in landfills and happen to be your worms' favorite food.

In a 2016 study published in Science Direct, it was concluded that in comparison to typical composting methods, "vermicomposting significantly reduced nitrogen loss by 10–20%, methane by 22–26% and nitrous oxide emissions by 26–36%." (Nigussie et al., 2016) These are not insignificant numbers. If we can repurpose some of our waste and help the environment at the same time, everyone benefits.

DID YOU KNOW?

Landfills

Vermicomposting can significantly reduce the amount of waste that you send to the landfill. Some experts estimate that home vermicomposting can reduce household waste by up to 70 percent. According to the EPA, the average American produces 4.9 pounds of waste per day.

Types of Worms and Worm Farms

Your choice of worms and type of worm farm system is highly dependent on your overall goals, the amount of space available for your vermicomposting project, and the time and attention you can invest in your new pets.

TYPES OF WORMS

There are thousands of species of earthworms, but only a handful can be used in vermiculture, so it's important that you choose a species that will be suited to this task. Of the several types of worms that can be used in vermicomposting, red worms and nightcrawlers are the most popular. **Hobby Worm Farm's Ken Mitzel** always uses red wigglers, as they are the easiest to raise and one of the best composters, producing sufficient amounts of castings. The castings are the waste or organic matter produced by the worms. The more castings your worms produce, the more material you will have to improve your soil and your crops.

Worms are classified into three ecological groups according to their behavior and needs. For vermicomposting purposes, only epigeic worms will work, as their natural habitat and tendencies are in tandem with the conditions that also make vermicomposting possible.

Do you want to start a worm farm solely for vermicomposting, or does vermiculture interest you as well? If you are considering expanding your hobby or operation into the cultivation or sale of worms, European or African nightcrawlers could be the better choice. Their size is bigger than the red wiggler, making them great fishing worms. In a tropical climate, you would do well to consider the African nightcrawler, as it eats quickly and generates a great number of castings.

Eisenia Fetida or Red Wigglers

Eisenia fetida, commonly known as the red worm or compost worm, is an especially tolerant and adaptable species that thrives in differing temperatures, farm types, and bedding. Red wigglers are also quick breeders and not prone to escapism like their earthworm counterparts.

Their bodies are reddish brown in color, with small rings around their body and a yellow tail. They generally tolerate temperatures between 55 and 80°F but will decrease their food consumption and breeding rates as temperatures get lower (Sherman, 2022). It's best to maintain the worm bed temperature somewhere around 70°F to keep them happy.

Red wigglers are native to Europe but are now found across the world and commonly used in vermicomposting by both individuals and agricultural producers. They are sold online by worm growers and can be purchased in large quantities. Start with at least 500 to 1000 worms, which is about one pound in weight, depending on the size and maturity of the worms.

Nightcrawlers or Earthworms

Both European nightcrawlers and African nightcrawlers can be used in vermicomposting as well. These two types of nightcrawlers have their own temperature requirements, with the European nightcrawler being more flexible. Choose carefully depending on your location and whether you plan to keep your worm farm outdoors or indoors.

The African nightcrawler prefers warmer temperatures, around 75°F, but can also tolerate higher temperatures of up to 95°F. At the same time, the African nightcrawler is quite intolerant to cold temperatures and will die if exposed to the cold long enough. The African nightcrawler would do well in a warm, even tropical climate.

On the other hand, the European nightcrawler can tolerate temperatures as low as 40°F, making it a much better choice for cooler climates. They will also survive temperatures up to around 80°F.

Although both types of nightcrawlers are considered epigeic and will mostly feed on the surface of the soil, the nightcrawler varieties will burrow deeper than the red wigglers. This creates some benefits, such as natural soil aeration, and is the reason these types of worms are sometimes used in plant beds.

Both types of nightcrawlers grow to much larger sizes than the red wiggler and would therefore be suitable for a vermiculture operation that requires more versatility. If, for example, your goal is to get rid of kitchen waste, produce vermicompost, and have a steady supply of fishing worms that are more substantial in size than the red wiggler, one of these two nightcrawler varieties would make a great choice.

If you are only interested in vermicomposting, red wigglers are your best choice, as they are easily available in most parts of the world. Their hardiness and tolerance make them a great choice for both amateur and seasoned worm farmers. If your first attempt at vermicomposting doesn't go as planned, it will be easy to replace them and start anew. Even better, they may be hardy enough to survive your first attempt, as ours were. After several flooding incidents in an especially heavy rainy season, our red worms persevered. We were glad we'd chosen them. Our African night-crawlers, on the other hand, made so many attempts to escape we started to feel like reluctant prison guards.

> "The prospective composter needs to decide if they just want composting worms or worms that can be used for bait as well. I look to buy worms locally when possible; if none are available, I research what suppliers have a good track record of shipping and delivering high-quality worms."
>
> **Kelly Hammel**
> Worm Hippie Worm Farm

THREE MOST POPULAR WORMS USED FOR VERMICOMPOSTING

**Eisenia Fetida,
Red Wiggler**

**Eudrilus Augeniae,
African Nightcrawler**

**Eisenia Hortensis,
European Nightcrawler**

CHARACTERISTICS

- Between a pinkish brown to a darker reddish brown in color
- 3–4 inches in length
- Stripey pattern alongside the body
- Tail is yellow or cream colored
- Visible saddle or clitellum
- Banding visible when stretched out
- Strong appetite and reproduction rates
- Reproduce quickly
- Tolerate a temperate climate

- Dark purple and gray coloring
- Can grow to about 8–10 inches in length—but is a bit thinner than Eisenia Hortensis
- Visible saddle or clitellum
- Banding visible when stretched out
- Reproduce even faster than red wigglers due to fast maturity rate
- Voracious eaters, producing more castings than all other compost worms!
- Tolerate hotter climates

- Deep red or purplish color, similar to red wigglers
- Can be between 4–8 inches long and as thick as a pencil
- Visible saddle or clitellum
- Banding visible when stretched out
- Slower rate of reproduction
- Eat slower than red wigglers and African nightcrawlers, so produce castings at a slower rate
- Tolerate cooler climates

THREE BASIC TYPES OF WORMS

Living conditions	Food	Size and Reproduction Rates
Anecic		
Burrowing worms that live in deeper layers of soil and don't spend much time at the surface of the earth. These types of worms build elaborate, vertical burrows. Example: the common earthworm or Canadian crawler	These worms come out of the earth to find food, such as leaves and other organic matter, and drag it into their burrows to eat.	They are larger than the other types of worms, generally between 8–15 cm long. Relatively low reproduction rates and long life cycles
Endogenic		
Burrowing worms as well, but live in more shallow layers of the earth, creating horizontal burrows. Example: the green earthworm or Allolobophora Chrorotica	Endogenic worms eat organic matter already found in the soil, so they rarely need to come up to the surface.	These worms are generally between 2–12 cm long and lack skin pigmentation. Lower rates of reproduction than epigeic worms and longer life cycles
Epigeic		
Epigeic worms do not build burrows and instead prefer to live close to the surface of the soil, where they have easy access to their preferred source of food. Example: red wiggler or Eisenia fetida, African and European nightcrawlers	They feed on leaf litter, decaying organic matter, and animal waste found on the surface of the soil.	They are the smallest of the three types, between 1–7 cm. High rate of reproduction, with a short life cycle

Note: Adapted from "Earthworms and Vermicomposting," by J. Domínguez in Sajal Ray (Ed.), *Earthworms — The Ecological Engineers of Soil* (2018), InTechOpen. https://www.intechopen.com/chapters/60445

TYPES OF WORM FARMS

The two most popular types of worm farm systems are the bin system and the flow-through system. Your choice of either system should be based on several factors, including the following:

- How much space you have available for your worm farm
- Whether you will be producing vermicompost solely for yourself or your family or for sale or agricultural use
- Outdoor or indoor setup
- How much you can invest in your initial setup
- How much time you can dedicate to the cleaning and overall maintenance of your worm farm

Bin Systems

Bin systems are the first and best choice for many beginners, as they are extremely easy to set up and can be established in a small space and without a large financial investment.

EXAMPLE OF BIN SYSTEM, CONCRETE CONTAINER OR BATHTUB

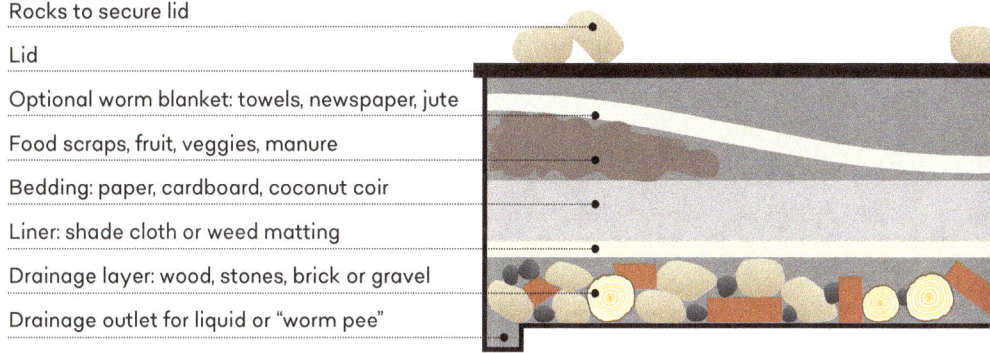

Rocks to secure lid

Lid

Optional worm blanket: towels, newspaper, jute

Food scraps, fruit, veggies, manure

Bedding: paper, cardboard, coconut coir

Liner: shade cloth or weed matting

Drainage layer: wood, stones, brick or gravel

Drainage outlet for liquid or "worm pee"

They can be as small as a plastic bin with holes in the bottom for aeration and a tray underneath, or they can be as large as a bathtub or an even larger concrete tank. Here in Nicaragua, even larger-scale worm farm operations are seen in the form of in-ground concrete structures. The low cost of concrete and relative ease of care are the primary reasons they are so popular.

Some examples of materials that can be used to start a worm farm in a bin system include plastic containers, pails or buckets, bathtubs, wooden boxes, and garbage cans.

Advantages of Bin Systems	Disadvantages of Bin Systems
Low cost to set up	Difficult to use in larger operations
Easy to set up for beginners	Poor oxygen and moisture control
Fit in compact, urban space	Bad smells
Can be made or purchased in a variety of sizes	Can be difficult and time-consuming to maintain and harvest
	Proliferation of pests

Flow-Through Systems

Flow-through systems are often used for larger-scale worm farms, although many compact flow-through systems are available and can be used in small spaces.

Flow-through vermicomposting systems can be constructed from a variety of materials, such as canvas or heavy-duty fabric, pails or garbage bins, and wooden boxes. Although it's possible to build your own flow-through system, it's a more complex project than making a simple bin system. If it's within your budget, it's best to start with one of the more popular fabric systems available for purchase or a ready-made commercial flow-through system. However, if you are feeling particularly adventurous, there are many designs available online and detailed instructions on YouTube that can help you build your own system from scratch.

Advantages of Flow-Through Systems	Disadvantages of Flow-Through Systems
➕ Can produce much larger amounts of vermicompost	➖ High up-front cost
➕ More efficiency in large operations	➖ Difficult to fit in small spaces
➕ Ideal oxygen and moisture control can be obtained.	➖ Some flow-through systems can dry out quickly due to high airflow (e.g., those made of fabric).
➕ Saves time on maintenance and harvest	➖ Larger models are bulky and difficult to move from place to place
➕ Easier to control pests	

2

SETTING UP YOUR
WORM FARM

Starting with a Solid Plan

Starting your first worm farm can be an exciting and fun project. Although easy and quick to start, a solid plan can go a long way toward helping you have a successful experience with minimal problems.

Before you buy your first batch of worms, pay attention to a few key considerations, including how much time and space you can dedicate to your worm farm, where your worm farm will be located, and the initial financial investment you are able to make. These factors will determine the style and size of the container you choose, the materials you will use, and the overall design and plans of your worm farm.

CONTAINERS

The size and style of your container will depend on the amount of organic waste and space you have available. Elise Pickett from the Urban Harvest suggests that having a steady and large enough food supply for your worms is important. They eat more than we might initially think. If you only have a small handful of scraps every couple of days, you probably won't be able to keep up with the demand. As we live on a fairly large farm, this is not a problem for our family. We have plenty of organic food scraps, mostly in the form of fruit peels, vegetable scraps, and yard waste. When we considered our first worm farm container, we knew it had to be large enough to handle the amount of waste our farm produced. A plastic bin under the counter would certainly not have sufficed. Thus, we started with two large wooden boxes we built ourselves, approximately six feet long by three feet wide.

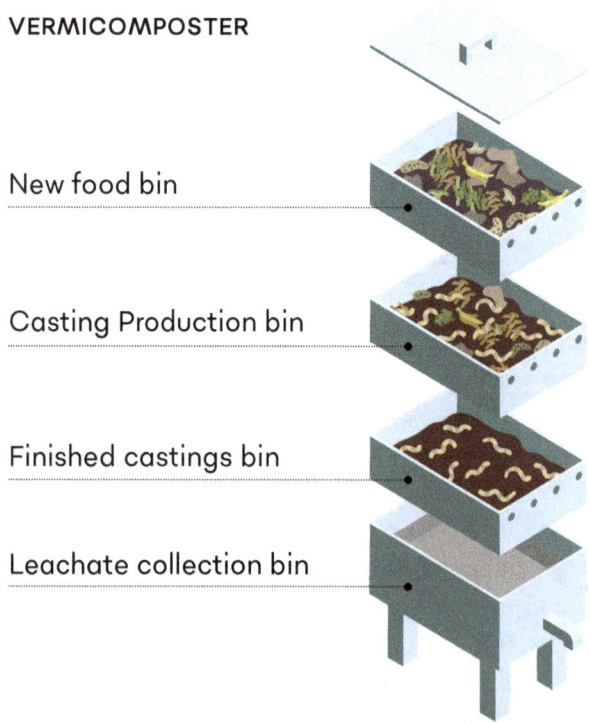

VERMICOMPOSTER

New food bin

Casting Production bin

Finished castings bin

Leachate collection bin

Most plastic bins or wooden boxes can be used as a container, and you can choose a size that is suitable for your needs. If you are starting your worm farm indoors, it will be easiest to start with either a single plastic bin, a stackable bin or tray system, a vertical worm tower, or a compact flow-through system that will fit in the space you have available.

If you will be producing vermicompost for sale or agricultural use, your best choices are large wooden containers, in or above-ground concrete bins, or large flow-through systems that allow you to house at least a few thousand worms and harvest a large number of castings.

Whichever type of container you choose, one of the most important considerations is that your container has proper ventilation and drainage to provide the worms with a suitable environment.

Voracious Appetite

Worms have a voracious appetite and can eat half their body weight each day; this means one pound of worms can eat half a pound of kitchen scraps daily. However, a new vermicomposting bin may not be able to process the same amount of waste as an established bin. Overfeeding is a common mistake and can result in a foul smell.

COMPARISON OF CONTAINER TYPES FOR YOUR WORM BIN

Type of Container	Pros	Cons
Plastic bin 	Low cost, variety of sizes, easy to maneuver, stackable, easy to make drainage holes	Poor airflow, higher temperature and moisture levels, labor-intensive, messy, low capacity
Plastic bucket 	Low cost, variety of sizes, easy to maneuver, compact shape, easy to make drainage holes, better drainage than plastic	Poor airflow, higher temperature and moisture levels, labor-intensive, messy, low capacity
Wooden container 	Breathable and absorbent, better airflow, more aesthetically pleasing than plastic, better insulation than plastic, higher capacity	Will not last as long as plastic, may be treated with chemicals, heavier than plastic, can suffer from lack of moisture
Bathtub 	Low cost (possibly free if you find an old bathtub), large size, built-in drainage, high capacity	Metal holds heat and moisture, poor airflow, labor-intensive to harvest, can be messy

Type of Container	Pros	Cons
Concrete container 	Relatively low cost, can be built to custom size, high capacity	Concrete holds heat and moisture, poor airflow, labor-intensive to harvest, can be messy
Fabric flow-through system 	Great airflow, easy to harvest, simple setup, high capacity	Might dry out due to high airflow, high cost
Flow-through system 	Simple to use, easy to harvest, excellent airflow, high capacity	Higher moisture due to increased airflow, high cost

"

I use a bin system for vermicomposting. It's easy, inexpensive, and fits easily underneath my seed-starting tables. It does require that I pay more attention to the worms to avoid excess moisture buildup, but that just gets me closer to the wonder of vermicomposting!"

Scott Wilson
Gardener Scott

MATERIALS

To start, you will need bedding material, such as shredded newspaper or coconut coir, and organic waste to feed the worms. To save on the cost of setting up your worm farm, try to use materials that are easily found in your own environment and everyday life. If you have plenty of dry leaves available, use them! If you have access to a lot of cardboard or shredded newspaper, use those instead.

In a study on the vermicomposting of banana leaf waste in India, it was concluded that vermicomposting is "a sustainable and economically viable method for treating banana leaf waste" (Mathew, 2018). A series of tests determined that combining banana leaf waste (either raw or dry) with cow dung produced quality compost and, more importantly, reduced India's ongoing problem of having great quantities of banana leaf waste littering its streets.

Here in Nicaragua, we have tremendous amounts of banana leaf waste as well. Our own farm has over three acres of Guineo bananas and plenty of banana leaves to spare. Naturally, our worm farm is heavily fed by this leaf waste, and it greatly helps us control the amount of waste we need to dispose of.

What we are trying to accomplish with the material is to add bulk and aeration to our worm farms, as well as provide our worms with a steady source of carbon. Other materials that can be used are straw or hay, aged manure, peat moss, and wood chips.

Remember that your worm bedding should be chemical-free and not contain any other materials that can harm your worms. This bedding will also act as a food source, so it needs to be clean and safe.

When choosing materials for the container you will use, keep in mind that metal and concrete structures will retain heat more than wood. If your containers are large enough, the worms can self-regulate by gravitating away from the sides of the container, but it's an important consideration in climates at the top or bottom end of the acceptable temperature spectrum.

In colder climates, a worm blanket or a worm condominium provides your worms with additional insulation and protection from the elements. Worm blankets placed on top of your container work both as insulation and to create the type of environment your worms thrive in: relatively moist, aerated, and dark.

> *Bedding is something you will use a lot more of than you think. When you start a farm, it should be between half to three-fourths filled with wetted-down bedding where the worms will live. We like to use non-glossy paper and cardboard shreds for bedding, but you can also use coconut coir, cow manure, or peat moss."*

Sam Evans
Sevans Wormery

MATERIALS THAT CAN BE USED FOR BEDDING IN YOUR WORM BIN

Bedding Materials		Preparation
Dry leaves		Let leaves dry outside first. Crush them down into smaller pieces if possible. Moisten and add to the worm bin. Use with other types of bedding that are more absorbent, such as cardboard or newspaper.
Straw		Recommended to use in conjunction with a more porous type of bedding that retains water. Break down into smaller pieces. Moisten and add to worm bed.
Hay		Recommended to use in conjunction with a more porous type of bedding that retains water. Break down into smaller pieces. Moisten with water before adding.
Coconut coir		Soak in water for a few hours before using, as it will make it easier to handle.
Aged manure		Ensure manure is aged before use. It may not require additional water.

Bedding Materials		Preparation
Shredded newspaper		Shred into small pieces. Saturate in water when starting the worm bed. Afterward, it can be added without water.
Shredded cardboard		Shred into small pieces. Add water when starting the worm bed. Afterward, it can be added without water.
Peat moss		Soak it in water for several hours, drain, and add to worm bin.
Wood chips		Use untreated wood chips only. Remove any large pieces that will take too long to break down. Moisten and add to worm bin.

DESIGN AND PLANS

There are many different designs and plans available for building or buying a worm farm. You can find plans online or purchase a ready-made kit. You can also get creative and build your own design using recycled materials.

You will have to consider the footprint of your worm farm project, the temperature fluctuations of your location, and the amount of time and effort you will be able to put into harvesting the castings. While you might find the idea of a large horizontal container interesting, if you don't have the space necessary, another type of design will have to suffice. In small apartments or small outdoor spaces, vertical or stacking systems work best, as they provide you with the greatest number of castings in the smallest possible space. You can fit thousands of worms in a vertical multi-bin system!

Time is also an important consideration. Harvesting castings from a bin system will inevitably prove more time-consuming than harvesting out of a flow-through system. Even if you only have a few feet of space available, a compact flow-through farm can be built or purchased to fit your space. Some of these systems are almost completely hands-off, requiring only a minimal amount of care. Yes, they are a bigger investment initially, but they pay you back in their usefulness.

The first thing to consider is where the worm farm is going to live. Will it be in a temperature-controlled environment, like a basement or garage, or will it live outside, where it will be susceptible to heat and cold? You want to pick a place where it will be easy to access and maintain. The other thing to consider is what kind of worm farm you want to get. There are a variety of sizes and types, so pick the one that will work best for your space. Stackable worm farms are great for smaller spaces because they can utilize vertical space, while continuous-flow bins are easier to use but are larger and take up more room."

Sam Evans
Sevans Wormery

LOCATION

Choose a location for your worm farm that is convenient and easy to access but also has the right temperature and moisture levels. William Clark of Moose Hill Worm Farm recounts that most people he talked to who failed at their worm farm had it where it would be out of sight and out of mind. This inevitably leads to failure of moisture control and overheating and freezing, leading to the death of worms. So, choose a location you will pass by daily, so you can peek in and take a look at what your worm friends are up to. Often, you can detect problems just by the scent of your worm farm, so being in proximity is useful.

You can place the container in a garage, basement, or another protected area, or you can set it up outdoors in a shaded location. If your worm farm will be outdoors, make sure that your container has a roof if you don't have a shaded spot available.

You will also want to ensure that your container has a secure lid, as birds and chickens are big fans. A roof doesn't have to fit perfectly so long as it stays in place. Often, I've seen concrete bins covered with a large piece of metal roof that is held in place by heavy rocks. It's not perfect, but it does the job well, and this type of roof is easy to maneuver on and off.

Most composting worms prefer a temperature range of 55–80°F and a moisture level of around 70%. While you consider the temperature, also make sure that your outdoor worm farm is not built in an area that is prone to flooding or where water will pool for a long period of time. Our family's first attempt at worm farming was heavily affected by a heavier-than-normal rainy season. We had chosen a location that seemed perfectly fine with a little bit of rain but later proved to be unmanageable when it flooded repeatedly. Our worms eventually started to wander off in search of drier pastures. We carefully considered the location of our second worm farm and built our next containers in an area that would never flood.

CHOOSING A LOCATION FOR YOUR VERMICOMPOST BIN DEPENDING ON THE CLIMATE

WARM

- Shady spot or north side of house, where you would get the least amount of midday sun

- Basement, if available

- On covered porch or under an awning

- Breezy spot

- Indoors

COLD

- Sunny spot or on the south side of house, where you would get the most amount of midday sun

- Garage

- Shed

- In an "insulation fort" surrounded by bags of hay or dry leaves

- Indoors

Setting It Up, Step-By-Step

Once you have planned out your worm farm, you can start setting it up by following these steps:

1. Acquire the container you will be using and choose a spot where your worms will be out of direct sunlight. Ideally, choose a location where it will be easy to control the temperature and moisture levels.

2. Cover your bin with a lid and make sure your container has drainage holes. Either build your own container or buy one that has the size requirements for the amount of vermicompost you want to produce.

3. Add bedding material to the container. Moisten the bedding material with water and add it to the container. The bedding should be about six inches deep and should take up between a third to one-half of your container's height before being packed down.

4. Add your worms. You can purchase worms from a local supplier or online. Red worms or nightcrawlers can be used, but red worms are generally easier to care for, easier to replace, and more suitable for vermicomposting.

I would highly suggest purchasing tools—such as a PH gauge, moisture meter, and thermometer—to help monitor those critical environmental elements. It's important to check those often in order to provide the best possible environment for the worms. After all, happy worms produce more cocoons and more castings!"

Ken Mitzel
Hobby Worm Farm

5. Feed your worms by adding organic waste to the container. You can add a small amount to the container every few days. The worms will consume the waste and turn it into compost.

6. Maintain the proper temperature and moisture levels. You can regulate the temperature by placing the container in a location that stays within the ideal range, and you can adjust the moisture level by adding water or bedding material as needed.

7. Correct any small issues that may come up. Keep an eye on your worm farm to make sure critters such as flies, mites, or ants don't invade your worms' habitat. If the organic matter is not being consumed quickly enough, reduce the amount, or chop it up into finer pieces. This will decrease the likelihood of it rotting and causing further trouble.

8. Harvest your vermicompost. Depending on the type of container you have chosen, you will need to empty the worm castings and refresh your farm with fresh bedding. Castings should be harvested anywhere from one to six months, depending on how many worms you have, the type of worm farm system you are using, and how much organic matter your worms are consuming.

FUN FACT

Common Worms

The most common worm used for vermicomposting is the red wiggler (Eisenia fetida.) While around 7,000 species of earthworms exist, only about seven are suitable for vermicomposting. In addition to red wigglers, common vermicomposting worms include the European nightcrawler, Indian or Malaysian blue, and African nightcrawler.

TIPS FOR MAINTAINING PROPER TEMPERATURE AND MOISTURE LEVELS

Maintaining consistent temperature and moisture levels in your container is very important to your vermiculture farm. Without it, your farm can turn from a fun project into a soggy or smelly mess. Fortunately, it's easier than it sounds, and the more you observe your worm farm, the easier it is to detect problems before they cause lasting damage.

To get precise measurements of your farm's moisture and temperature levels, you can purchase a thermometer and a moisture meter. Both are usually sold by worm growers online.

Alternatively, you can use your nose and a basic "squeeze test" to detect temperature and moisture problems. The basic rule is that if you squeeze a handful of the bedding and vermicompost in your hand and only a few drops come out, your farm has the correct level of moisture. Any more than that, and there is probably an excess; any less, and your bedding needs more moisture.

An unpleasant odor usually indicates anaerobic conditions, either from high temperatures and lack of air circulation or excess moisture.

FIXING TEMPERATURE AND MOISTURE LEVELS

Temperature		Moisture	
Too hot	**Too cold**	**Too wet**	**Too dry**
• Move to the shade	• Bring indoors	• Add dry bedding	• Add wet bedding
• Bring indoors	• Keep a lid on	• Feed less or lower moisture foods	• Add high-moisture foods
• Install a fan	• Add insulation	• Check drainage	• Mist top of container with spray bottle
• Add moisture to lower temperature	• Install a heater	• Remove visible moisture	
		• Increase drainage	

3

FEEDING YOUR WORM FARM

What to Feed Your Worms

Worms are omnivores and will consume a wide range of organic materials. Audrey Wynkoop of the Worm Bucket advises that one of the easiest ways to source organic waste for your worm farm is to save your own kitchen scraps, such as fruit and vegetable peels, coffee grounds, and tea leaves. Your worms can also eat eggshells, pulp from your juicer, and aged or composted manure.

They will also happily consume the bedding you use in your containers, although it will be broken down at a much slower rate than the food. You'll notice in the chart below that bedding has a much higher carbon content, while things like food and fruit and vegetable scraps have a much higher nitrogen content.

The ideal carbon-to-nitrogen ratio in your worm bin is around 50:1 or higher. Anything lower than that, and you will notice that your bin will increase in moisture and start to attract common pests and perhaps even predators. This anaerobic environment will also increase the temperature in your container. Keeping the nitrogen in check by adding high-carbon sources will keep your worms in good health and your bin problem-free.

As you think about what type of food you will provide your worms, remember that, in this case, input has a great degree of control over output. When you feed your worms a variety of quality foods of differing nutritional, mineral, and micronutrient content, your vermicompost will reflect that. You will, in turn, have a source of vermicompost that will feed your garden and help it to flourish at its best.

DIFFERENT TYPES OF WORM FOOD AND THEIR CARBON TO NITROGEN RATIOS

Food Source	Carbon To Nitrogen Ratio
Cardboard, Shredded	350:1
Newspaper, Shredded	175:1
Fruit Waste	35:1
Vegetable Scraps	25:1
Food Scraps	20:1
Leaves	60:1
Straw	75:1
Wood Chips	400:1
Hay	25:1
Manure	15:1
Garden Waste	30:1
Coffee Grounds	20:1

FRUITS AND VEGETABLES

Fresh or cooked fruits and vegetables will probably make up a substantial source of your worms' food intake. They make excellent food for worms, as they provide a balanced diet and are rich in nutrients.

Worms tend to favor soft fruit, especially anything in the Cucurbitaceae family. Although not an extensive list, here are some of the fruits your worms will enjoy:

- Papaya
- Honeydew
- Apples and apple cores
- Strawberries
- Watermelon
- Avocados
- Peaches
- Apricots
- Cantaloupe
- Bananas
- Pumpkin
- Squash

Vegetables are also high on the list of foods worms prefer, starting with leafy greens like Swiss chard, lettuce, and kale. These are easy for your worms to consume and break down quickly. Your worms will also eat other vegetables like broccoli, cauliflower, potato skins, carrot peels, and cabbage leaves. The list of vegetables you can feed your worms is long, so have fun with it!

Avoid feeding your worms citrus fruits, onions, and garlic, as these can be harmful to them.

> *I feed them all my fruit scraps—especially bananas and papayas!— lots of leafy greens from my home and local organic grocery store, and some occasional coffee grounds. I like to save my fruit and veggie scraps in the fridge or freezer until it's feeding time."*

Elise Pickett
The Urban Gardener

COFFEE GROUNDS

If you've been using your green thumb long enough, you've probably heard that coffee grounds are a useful addition to your composting heap and your garden. Your worm farm will also benefit from coffee grounds since they provide a great source of nitrogen for your worms.

You can add your coffee grounds to the bedding, or you can add them at feeding time with other sources of food. Either way, make sure the grounds are moist, as this will also help maintain the moisture balance in your container. You can even add a little bit of liquid coffee to your bedding, as it will provide some of the same benefits when your worms eat it.

If you're not an avid coffee drinker, you can ask your local coffee shop to save you some of their used coffee grounds. You will be helping them repurpose some of their trash and offering a varied diet to your worms. Coffee grounds also help to balance the pH of the compost. Just make sure not to rely on them too heavily, as this will make your farm's environment too acidic.

TEA LEAVES AND BAGS

Tea leaves and bags are a good source of nitrogen for your worms. They also help to balance the pH of the compost.

Ensure that if you are feeding your worms tea bags, they are not a brand that contains plastic, as some tea bags do. Check the packaging to make sure your tea bags are biodegradable. Your worms will also consume the paper tab and string, but these will break down slower than the tea leaves.

EGGSHELLS

Eggshells are a good source of calcium for your worms and can help balance the compost's pH. They are often the easiest fix if you have overindulged your worms with coffee grounds. Adding eggshells will effectively neutralize the pH of your worm bin if the environment has become too acidic.

Like other small animals and birds, worms use their gizzards to break down other particles of food, and coffee grounds and eggshells are a good source of the grit that is necessary to their digestive processes.

Crush the shells into small pieces before adding them to your worm farm. Otherwise, they will take a long time to break down and get in the way of your worms moving around the bin freely. Your worms cannot break the eggshells down themselves, as they lack teeth. You can also add your eggshells as a powder for faster consumption.

Eggshells are also believed to be an aphrodisiac for your worms and raise reproduction rates due to their calcium content. Just as with coffee grounds, don't add too many at once or too often. A small handful of crushed or powdered eggshells a few times a month is plenty.

CARDBOARD

Cardboard is a good source of carbon for your worms and can also help to balance the pH of the compost. Shred the cardboard into small pieces before adding it to your worm farm.

While your worms cannot live solely on cardboard, the cardboard can be mixed into other foods, such as kitchen scraps. Being one of the higher carbon foods or bedding you can add to your vermicompost bin, cardboard allows you to also feed foods that are very high in nitrogen without having to worry about disturbing the overall carbon-to-nitrogen ratio.

In highly developed parts of the world, cardboard is readily available in the form of packing materials, packaged foods such as cereals and candies, and egg cartons. Here in Nicaragua, cardboard is harder to source as product deliveries are not common, and there is much more reliance on fresh and local foods without packaging. However, as suggested earlier, use what you have on hand. If cardboard is plentiful where you are, your worms will be happy to eat it.

DID YOU KNOW?

Cellulase

Scientists have recently discovered that earthworms possess a unique enzyme in their digestive tracts, which allows them to break down cellulose. This cellulose-devouring gut enzyme, called cellulase, enables the worms to break cellulose down into carbon, hydrogen, and oxygen.

What to Avoid Feeding Your Worms

Avoid feeding your worms meat, dairy products, sweets, and other processed foods, as these can attract pests and create an unpleasant smell. Kitchen waste that breaks down very slowly, such as bones or eggshells that have not been crushed, is also not recommended.

If you will be adding yard waste to your worm farm, make sure it's free of pesticides and other chemicals that can harm your worms.

Other foods your worms will not enjoy are citrus fruits, nonbiodegradable items such as tea bags containing plastics, colored or glossy paper, and spicy foods.

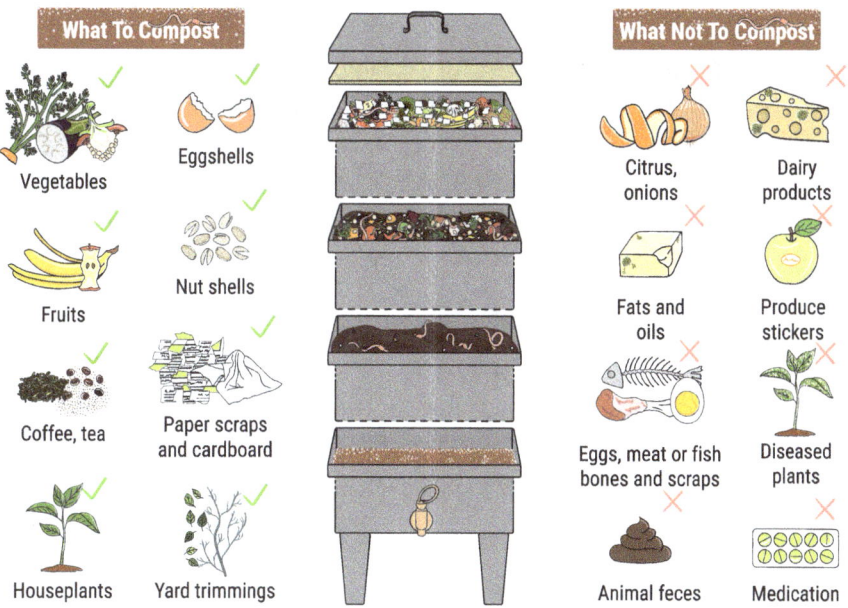

What To Compost

Vegetables

Eggshells

Fruits

Nut shells

Coffee, tea

Paper scraps and cardboard

Houseplants

Yard trimmings

What Not To Compost

Citrus, onions

Dairy products

Fats and oils

Produce stickers

Eggs, meat or fish bones and scraps

Diseased plants

Animal feces

Medication

How Much to Feed Your Worms

It's important not to overfeed your worms, as this can lead to an excess of uneaten food and a buildup of harmful bacteria. A general rule of thumb is to feed your worms an amount that is equal to about half their body weight per day. For example, if you have one pound of worms, you should feed them about 0.5 pounds of food per day.

If your container has a large amount of surface area, it's best to place the food in a thin layer rather than piling it all up, as this will make it easier for the worms to eat before it starts to decompose. While your worms cannot handle large amounts of food at one time, they will steadily grow in numbers as long as you feed them the right amount of food.

Daily feedings are not necessary. Calculate the proper amount of food for your farm and then modify their feeding schedule to what works best for you and your family. **Audrey, who writes the Worm Bucket**, only feeds her worms every seven to ten days and also avoids things like potato peels and uncooked carrots that take a long time to break down. You can leave your worm farm for a week or two while you go on vacation and not worry that you will come back to a disaster. If you must question whether or not to feed them more, lean toward no.

If you notice that your worms are not eating all the food you are providing, reduce the amount of food you are giving them. An excess of uneaten food or an unpleasant smell is a certain sign that you are overfeeding your worms.

Avoiding Common Feeding Mistakes

Feeding your worms is where most beginner mistakes happen. It's easy to overfeed them, feed them the wrong foods, or do it in a way that leads to problems with moisture and pests.

OVERFEEDING

As mentioned above, it's important not to overfeed your worms. This can lead to an excess of uneaten food and a buildup of harmful bacteria.

In the beginning of your vermicomposting project, you may find yourself hovering around your worm bin with some degree of frequency and excitement, looking to see what your worm friends are doing and finding things you can do to make them happier and more productive. Your eagerness can easily lead to more frequent feedings than necessary, but your worms can only handle so much food at once!

Take it slowly and keep to a regular and measured feeding schedule so you can observe just how much food your worms can eat. Once you are familiar with their eating habits, you will have more flexibility in how and what you feed them. First, you must get to know them, and that's one of the fun parts of vermicomposting.

Don't overfeed the worms! When in doubt, add less food and give them time to keep working on the food you've already given them before adding more. More worms are killed by overfeeding than by underfeeding. They will eat the bedding and the castings if needed, and don't require daily feedings."

Erika Babbitt-Rogers
Wyoming Worm Wrangler

FEEDING THE WRONG MATERIALS

As noted earlier, avoid feeding your worms meat, dairy products, and processed foods, as these can attract pests and create an unpleasant smell.

In addition to the wrong foods, pay attention to the carbon and nitrogen content of various bedding materials and foods you are using. Part of maintaining your worm bin is keeping everything in proper balance, and this will come more easily with a little experience.

BURYING THE FOOD

It's important to place the food on the surface of the bedding, rather than burying it. This will allow the worms to easily access the food and prevent an excess of uneaten food from building up. The environment can quickly turn anaerobic if you bury the food scraps.

Since vermicompost worms feed at the surface of the soil, they will be coming up for food regularly, so you don't have to worry about the food being found. You are much more likely to find clumps of uneaten, decaying food if you bury it. Spreading the food in an even layer on the surface is also helpful and will make it easier for your worms to eat quickly.

Odorless Vermicomposting

When done correctly, vermicomposting is an odorless process. However, a foul-smelling vermicomposting bin can indicate an imbalance. The first step in addressing an imbalanced vermicompost bin is to aerate the system by lifting the top layer of waste and bedding because composting and worms require oxygen to break down waste. Overfeeding is one of the most common causes of a smelly worm farm.

Carbon to Nitrogen Ratios: An Important Detail

Maintaining the proper balance of carbon and nitrogen is one of the most important considerations in your worm farming adventure. "Vermicompost bins should have a carbon-to-nitrogen (C:N) ratio of around 50:1. That's significantly more carbon than regular compost bins" (Hendrickson, 2021). When you are starting your worm farm, give some thought to the combination of food and bedding you will be using. If you plan to feed a lot of nitrogen-rich foods like fruit peels and vegetable scraps, it's best to choose a bedding material higher in carbon, such as shredded cardboard or newspaper, so that the balance of nitrogen to carbon in your farm will be maintained in proper balance.

Kelly Hammel of Worm Hippie Worm Farm advises that food waste usually creates a worm bin that is too wet and then turns anaerobic and starts to smell, which in turn causes the worms to die. When starting out raising worms, only provide enough food for the worms to clean up within a week before adding more. When adding food waste, also add some carbon as well.

If, on the other hand, you don't have access to a lot of greens and instead plan to provide your worms with a lot of browns like corn stalks, straw and hay, eggshells, dried leaves, or grass, you can choose bedding that is lower in nitrogen content.

THE ROLE OF CARBON AND NITROGEN ON TEMPERATURE AND MOISTURE LEVELS OF YOUR WORM FARM

High Heat

High Moisture

High Nitrogen Input:
Food scraps, vegetables and fruits, coffee grounds, fresh grass and weeds, aged manure

High Carbon Input:
Dry leaves, straw and hay, wood chips or twigs, shredded cardboard and newspaper, dry leaves

Low Heat

Low Moisture

4

HARVESTING
YOUR
VERMICOMPOST

Vermicompost Harvesting

After several months of vermicomposting, you will have a rich, nutrient-dense compost that is ready to be harvested. In this chapter, we'll cover the process of harvesting your vermicompost, including how to separate the compost from the worms and bedding and how to use the compost in your garden or landscape.

The process of separating the compost from the worms and bedding is called "worm casting." There are several options available to harvest your vermicompost, some requiring a lot of manual labor, some almost none, and a few that fall somewhere in between. All methods are relatively easy, and you can try several to see which best suits you. **David Trood, known as the Weedy Gardener**, uses a bathtub for his vermicompost farm and tells us that a week before he harvests castings, he moves all the food to one end of his bathtub. Over a few days, the worms will migrate toward the end of the tub that has food. Then, he is able to take the castings, leaving the worms at the other end of the tub. It's one of the simplest and most efficient methods, and it's one that my family and I have used ourselves with great success.

We started our vermicomposting adventures by manually separating our vermicompost, first with our own hands, then with the aid of some small rakes and a homemade sifting tray. After a few lengthy sessions, we started using lights and food to lure the worms where we wanted them to go while we removed their castings as gently and unobtrusively as we could. We find that using light or food makes things go much faster and keeps our worms happier as they congregate together and away from the castings we are trying to remove.

Audrey Wynkoop of the Worm Bucket also feels that the light method disturbs the worms as little as possible and recommends using a hand trowel to remove the top layer of castings into another container and repeating until you have harvested all the castings. Additionally, she points out that with this method you don't even have to touch the worms to harvest.

If you make or purchase a sifting screen that is fine enough, approximately one-eighth of an inch or less, you will also be able to sift most of the eggs or cocoons and return them to your container. We suggest you try a variety of harvesting methods, as they all have their benefits and downfalls. In the end, the best method is the one that you will be most comfortable using.

COMPARISON OF DIFFERENT HARVESTING METHODS

Harvesting Method	Pros	Cons
Manual sorting	Free. It can be disturbing to the worms but less so than a screening machine or other automatic methods.	It is laborious, time-consuming, and messy.
Bait or food migration method	Free. It requires less effort than manual sorting and is least disturbing to the worms.	It is slower than the light method and more time-consuming.

Using migration methods can be the best way to harvest worm castings with minimal impact on the worms. It can take longer than other methods, but providing a new, good source of food and allowing the worms to move toward it means you can avoid making physical contact with most of them when harvesting."

Scott Wilson
Gardener Scott

The light method	Free if you have a portable light source. It is much faster than manual sorting, as the worms will gather in piles and help the process.	It can still be time-consuming, and you will definitely get your hands dirty.

I've always used the light method. It's more labor intensive, but it does spare the worms. Worms do not like the light, so they bury themselves in the compost to get away from it, leaving you free to simply scrape the top layer off. They run, and you scrape until you have nothing left but worms!"

Ken Mitzel
Hobby Worm Farm

Harvesting Method	Pros	Cons
Screening method	It is quick and efficient—very hands-off.	It will disturb the worms the most; can even result in some deaths. Screening machines are pricey.
Stackable tray system	It is reasonably priced with not much labor involved.	It is hard to control moisture levels; some vermicompost may be wet and clumpy.

I use stacking trays, plastic totes, and a Worm Inn for indoor composting in my house. I like that the trays allow for a bit of growth without wasted space, and the Worm Inn provides a little easier access to the finished castings without having to disrupt the worms as much. Plastic totes are cheap but do require a little more knowledge to keep the environment ideal for the worms."

Erika Babbitt-Rogers
Wyoming Worm Wrangler

DID YOU KNOW?

Black Gold

Worm castings, sometimes called black gold by enthusiastic gardeners, are a garden soil superfood. Also known as vermicast, worm castings are the worms' excrement and are rich in micronutrients, such as magnesium, zinc, copper, calcium, iron, and sulfur.

Separating the compost from the worms and bedding

THE MANUAL METHOD

1. **Remove the top layer of bedding:** Using a shovel or rake, remove the top layer of bedding from the container. This will expose the worms and any uneaten food.

2. **Push the worms and uneaten food to one side:** Using a tool or your hands, gently push the worms and uneaten food to one side of the container.

3. **Remove the finished compost:** Using a shovel, rake, or your bare hands, remove the finished compost from the opposite side of the container. Be sure to leave the worms and uneaten food behind.

4. **Repeat the process:** Once you have removed all the finished compost, you can add a new layer of bedding and organic waste to the container and repeat the process.

5. **Sift the compost:** To remove any remaining bedding and worms, you can sift the compost through a fine-mesh screen. This will leave you with clean and worm-free compost that is ready to use.

THE BAIT OR FOOD MIGRATION METHOD

1. The bait or food migration method works best if you don't feed your worms for at least a week before harvest. This ensures that when you do feed them, they will be thrilled to gather exactly where you want them. Thus, **restricting food before harvest should be your first step and needs to be considered ahead of time.**

2. **Make a pile of vermicompost on one side of your container:** Making a pile will allow you to harvest the most compost at one time. You will continue to make piles until you have harvested your entire container.

3. **Place the food on one side or in a corner of your container:** The tastier the food, the better. Cantaloupe, pumpkin, squash, and anything in the Cucurbitaceae family that will break down quickly will be a big hit with your worms.

4. **Wait anywhere between 24 hours and one week, and then harvest your castings:** Wait time depends on how long before harvest you've been restricting food, what type of food you feed the worms, and the number of worms in your container. As your worms migrate to the part of your container that has the food, harvest the castings on the other side of the container.

5. **Sift the compost:** The last step, completely optional, is to sift your compost.

THE LIGHT METHOD

1. **As with the other harvesting methods, remove the top layer of bedding and food:** Use your hands, a rake, or a small shovel for this step. Your goal is to remove any large pieces of uneaten food as well as bedding.

2. **Make a pile of vermicompost on one side of your container:** Making a pile will allow you to harvest the most compost at one time. You will continue to make piles until you have harvested your entire container.

3. **Set up your light source over the first pile you've made:** Secure your light above the mound or pile, so you can use your hands or a small shovel to harvest the vermicompost.

4. **Remove the vermicompost:** As your worms move away from the light, use your hands or tools to remove the compost and put it in your collecting container.

5. **Repeat the process, making more piles:** Keep making new mounds of compost and allowing the worms to bury themselves deeper into the pile until all compost has been harvested.

6. **Sift the compost:** The last step is to sift your compost, which is always optional and may not be necessary if you are using it in your own garden rather than selling it.

Pest Control

Applying vermicompost to your garden beds can help reduce unwanted pests and diseases. For example, many gardeners attest that vermicompost can repel white flies, spider mites, and aphids more efficiently than commercial chemical pesticides.

STACKABLE TRAY SYSTEMS

1. **With a stackable tray system, your harvesting method is to add an additional tray to the top of your system:** Load the new tray with fresh bedding and food, and place it on top of your existing stackable tray system.

2. **Harvest the lowest tray in your stackable tray system:** Once your worms have migrated into the top tray that now holds all the fresh food, the bottom tray should be nothing but rich vermicompost. Remove the vermicompost.

3. **Sift the compost:** Use your sifting tray to separate any large chunks of bedding or unfinished food left in your compost.

4. **Keep the empty tray until you want to harvest more castings, and then repeat the process by placing this tray at the top of your system:** Repeat this process to have an ongoing supply of vermicompost.

FUN FACT

Vermicompost Tea

Vermicompost tea is a popular method for enriching your plants or garden with nutrients. This liquid fertilizer is made by steeping vermicompost in the water used for watering your plants. While it's a fantastic option for houseplants, purchasing it ready-made can be quite expensive, however making your own vermicompost tea can be cost-effective. Regardless, it remains one of the best fertilizers available for plant health.

Using Your Vermicompost

There are many ways to use your vermicompost in your garden or landscape. Here are a few ideas and ways we have used vermicompost ourselves and the ways it has helped our gardening endeavors.

USE IT AS A GENERAL SOIL AMENDMENT

Vermicompost is a rich, nutrient-dense soil amendment that can be used to improve the health and fertility of your soil. Simply mix it into your soil and use it to increase the health, yield, good bacteria content, and growth rate of just about anything you plant. Since vermicompost is completely natural, the microorganisms and nutrients it holds are readily available to be used by your plants. Your worm castings will bind to the soil and stay in the soil for longer than regular fertilizers, as well as provide natural aeration and porosity to the structure of your soil. What's more, since vermicompost is natural and mild-mannered, it won't burn your plants like regular compost, making it safe and easy to use for the novice gardener.

Preparing the soil for an extension of our veggie garden. Vermicompost used underneath the mulch to enrich the soil.

USE IT BENEATH MULCH

In our tropical climate, mulch is a necessary part of our gardening tactic. Without it, our plants cannot retain enough water, and the powerful heat weakens and even kills our young trees and plants before they have a chance to boost their strength and stand on their own. To reduce our costs, we use whatever mulch we have available on our property. In our case, it often takes the form of banana leaves and stalks, fresh grass or hay, tree leaves and wood shavings, or bits and pieces of fallen tree branches we chop up. We simply cut and place banana leaves around the base of our trees, and the banana stalks are cut or left whole and placed around our trees and plants to help provide and retain moisture.

Vermicompost can also be used beneath your mulch to help retain moisture in the soil and suppress weeds. Simply spread a thin layer of vermicompost under your mulch, and water it in to allow the nutrients to seep into the soil. For instance, we often place vermicompost under our banana leaves or stalks, ensuring it nourishes the soil as we water the plants.

You can also mix vermicompost into your cover crops, increasing the nutritive powers of your cover crops and enriching your soil substantially to ready it for the next season of production.

Mulch used at the base of plants and trees, with vermicompost underneath.

USE IT IN SEED BEDS

Vermicompost is a great option for seed beds, perfect for young seedlings in need of nutrition. Simply mix it into your seed rows, laying it around the roots of your seedlings or mixing it directly into the soil you are using to plant. Your seedlings will have a fertile source of food as they stretch out and make their debut in your garden.

Another way to help your seeds is by soaking them in vermicompost tea. Some greenhouse experiments were published by the American Society of Horticultural Sciences, in which tomato and lettuce seeds were soaked in vermicompost water or tea. It was concluded that vermicompost tea raised germination rates significantly, and the more concentrated the tea, the better it worked. (Arancon et al., 2012)

USE IT IN YOUR CONTAINER GARDENS OR HOUSEPLANTS

Vermicompost is a great option for container gardens and indoor plants. Simply mix it with potting soil or use it as the sole growing medium in your containers.

If you are using it as an amendment, you can remove a little of the soil in your containers and replace it with vermicompost. Do this all around the circumference of your potting containers so your plants readily absorb the nutrients. You can also sprinkle it on top of your houseplant containers as a top dressing, using between half an inch to an inch of vermicompost.

TIPS FOR HARVESTING VERMICOMPOST

- Wait until your worm castings are ready to harvest! Eager vermicomposters are excited about their first harvest, but your vermicompost won't be ready and at its full capacity of nutrition and richness until it is a deep, dark brown color.

- Don't get stuck on one harvesting method if it seems difficult. Try different methods until you find the one that fits your lifestyle and feels easiest.

- Don't be afraid to use tools to harvest your worm castings. Many of us get hung up on having to touch the worms with our hands or harvesting vermicompost in a way that won't disturb the worms at all. Your worms will be fine with a little excitement on harvest day, and tools were made to aid us in our work. Use gloves, small shovels, rakes, and anything else that helps you have fun and feel comfortable in the process.

- If none of the above methods work for you, there is always the option of using a harvester or screening machine, which you can build yourself or purchase. Prices start at around $1000 and go up from there, depending on their functionality and capacity.

5

COMMON
PROBLEMS

Even with proper care, it's common to encounter problems with your worm farm from time to time. In this chapter, we'll cover some common issues that can arise and offer tips on how to fix them. By troubleshooting these issues, you can keep your worm farm running smoothly and produce high-quality compost.

If you find yourself face to face with a smelly or pest-infested worm farm, don't despair. There are solutions to just about any vermicomposting problem, and if not, you always get another chance to start again and do better the second time.

The Most Common Problems in your Vermicompost Farm

The most common problems in vermicomposting are also the easiest to spot, and they are the easiest to troubleshoot.

They include the following:

- Overfeeding (Easily the most common problem among eager vermicompost beginners!)
- Feeding your worms foods that attract pests or harm the health of your worms
- Neglecting to rid your container of food or bedding that is not decomposing fast enough
- Not using any pest control methods
- Improper balance of bedding materials
- Insufficient moisture
- Overcrowding
- Too much moisture

DID YOU KNOW?

Regeneration

Many worms can not only survive being cut in half but can regenerate their missing parts. However, this is only true under certain conditions. For example, most worms can regenerate a lost tail or a partially amputated head. However, for red wigglers, the more head segments lost, the less likely they will be able to regenerate their head fully. This remarkable ability is caused by stem cells called neoblasts.

Resolving Your Worm Farm's Most Common Problems

Once your vermicompost bin is established and in good balance, it will house a variety of insects and organisms. This is a healthy and normal part of your vermicompost project. Our intention is to keep things in balance and functioning well together.

However, sometimes your farm's complex ecosystem can tip out of balance. It's then that you will encounter complications such as pests or poor-quality compost. Bringing this delicate ecosystem back into harmony will resolve most of these problems and return your worm farm to a comfortably balanced state.

PEST INFESTATIONS

One common issue that can arise with worm farms is pest infestation. Pests such as black soldier flies, earthworm mites, and fungus gnats can be attracted to the organic waste in your worm farm and can harm the worms or interfere with the composting process. To prevent and control pest infestations, follow the guidelines listed below.

As you go forward, keep in mind that when you are faced with pests in your vermicompost containers, your task is to tackle the underlying problem rather than trying to figure out how to get rid of the pests themselves. Of course, getting rid of them is what you'll do in the process, but if you can prevent the problem from reoccurring, you'll have a healthy and happy worm farm in the long run.

It's also important to distinguish between good pests and bad pests because only some pests need to be eradicated, while others are harmless or even beneficial. Some of the good pests include roly polies, earwigs, and millipedes. Millipedes are not to be confused with centipedes, which are one of the few worm predators you may encounter in your bins. Pluck them out with care, as some types have a mean bite.

Most mites are not directly harmful to your worms, at least not in small numbers, but they are unsightly and, if left to multiply, will compete for food with your worms. Earthworm mites are a different matter altogether, and a large infestation is generally considered a serious problem.

> *Fruit flies and fungus gnats are the bane of most worm farmers at some point or another. I have found that freezing my food scraps helps. I also try to wrap all food scraps in a piece of newspaper before burying it into the bedding of my bins. That way, the flies are less likely to get to the foodstuff, but the worms will still eat through the paper once it has rotted enough for them to consume it."*

Erika Babbitt-Rogers
Wyoming Worm Wrangler

OVERFEEDING YOUR WORMS

Overfeeding is the number one problem experienced by beginner vermicomposters. It's no wonder since it's the one thing we can freely "do" with our worms; it's our way of interacting with them. We can't pet them as we would a cat, and we can't take them for a walk like a dog, so we feed them tasty cantaloupe seeds and save juicy bits of watermelon and discarded salad greens for them, all in an effort to treat them well and make them happy.

David Trood concurs with this sentiment, saying, "The general mistake is putting too much food at once into the worm farm. Feed them a little, but regularly. That way, the food scraps don't go rotten before the worms have time to eat them."

Unfortunately, overfeeding can lead to a buildup of uneaten food, which can attract pests and increase the temperature in your container. It's important to feed your worms an appropriate amount. Common advice states that this is between 50 to 100% of their weight in organic waste, but it's often too much. If you are in doubt about the amount you should feed your worms, start with half that amount and use your own observational skills to gauge if the food is being consumed before it gets a chance to decompose. If it is, you are on the right track, and you can start increasing the amount a little at a time.

If the food is not being eaten, then it is safe to reduce the amount. Your worms can handle extended periods without food, and they are far more likely to suffer from overfeeding than underfeeding. Fruit flies and fungus gnats are notoriously attracted to fermented foods and commonly appear when you overfeed. They are also quite difficult to get rid of.

There isn't much that pests love more than uneaten food left in your worm container. You only want to feed your worms, not other visitors as well. Make sure that slow-to-decompose foods are removed before they start to ferment and attract other hungry guests.

FEEDING YOUR WORMS THE WRONG FOODS

Avoid feeding your worms meat, dairy products, and processed foods, as these can attract pests and create an unpleasant smell. Stick to feeding your worms organic fruits and vegetables, coffee grounds, and other suitable materials. In general, foods that decompose quickly will have less opportunity to attract pests as they are consumed quickly by your worms.

Bones, meat, and anything greasy and particularly tasty will quickly bring in rodents, vermin, and other pests that typically scavenge for food scraps. If you are on a farm, like us, or away from an urban area, you have a good chance of attracting animals that will not only ravage your worm bin in search of a tasty treat but may cause other problems as well. For example, feeding your worm farm discarded bones could be an excellent way to attract animals that will also try to get into your chicken coop.

If you find yourself running out of fresh organic fruit and veggie scraps, it's best to feed less rather than supplement with foods that will attract pests. Once insects or worms settle into your worm containers, it's a tedious process to get rid of them. Not impossible, but certainly laborious.

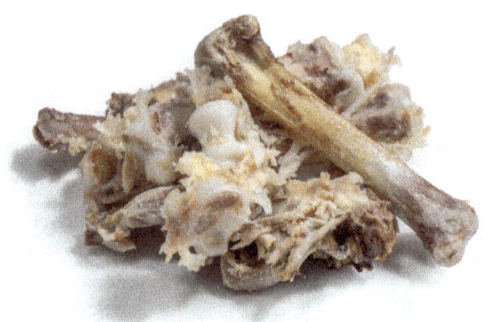

KEEP YOUR CONTAINER CLEAN

Regularly remove any uneaten food and discarded bedding from the container to prevent the buildup of harmful bacteria and attractants for pests.

For the most part, harvesting your compost bin goes hand in hand with cleaning it. While you are collecting your finished castings from your bin, you are also removing:

- Any food scraps that haven't decomposed fully, such as eggshells or pits
- Trash, such as bits of plastic bags or food stickers that have ended up in your bin
- Old bedding that hasn't decomposed or that is starting to ferment

As you harvest your vermicompost, removing any old bedding and scraps, your bin is refreshed and ready for a new cycle of production.

USE A PEST CONTROL METHOD

If you notice a pest infestation in your worm farm, you can use a natural pest control method, such as introducing predator nematodes, using a neem oil spray, or diatomaceous earth. Using screens and wire meshing to prevent entrance to black soldier flies and rodents would also be considered a pest control method. In fact, anything you do to prevent pests from settling into your bins will be part of pest control, and this includes keeping your bins in a balanced ecological state.

These are some common pest control methods:

NEEM

Neem oil spray, "cakes," or pellets can be purchased in some gardening centers or online. Alternatively, it's easy to make a spray at home if you have access to neem trees, as we do on our farm. Neem oil is effective against a variety of potential intruders to your worm farm and is especially effective against mites. Spray it on the surface of your bin or on any new bedding you are adding to it.

FOOD-GRADE DIATOMACEOUS EARTH

Diatomaceous earth is composed of the remains of single-celled marine plant organisms that have hard-shelled bodies. These remains are ground up and used in myriad ways, one of which is pest control. Although your worms can digest diatomaceous earth, it will cause other insects to die by drying their bodies of moisture and oils. They will essentially die of dehydration.

You can also sprinkle diatomaceous earth around the perimeter of your bins to prevent ants and other small insects from entering in the first place.

Make sure that you purchase only food-grade diatomaceous earth, as the other varieties are harmful to both animals and humans. You don't need to use a lot for pest control, as a little goes a long way.

INTRODUCTION OF PREDATOR NEMATODES

Packs of predatory nematodes can be purchased and applied to your worm bin to prevent fly and gnat infestations. These microscopic worms will find and kill fly and gnat maggots and other microscopic bugs, hopefully preventing their population from thriving in your bin. Once your vermicompost farm is well established, it will encourage the proliferation of nematodes and other beneficial organisms, so you won't have to rely on purchasing these packs indefinitely. They are very useful at the beginning when your farm hasn't established a powerful ecosystem yet, or if you have a particularly acute infestation of the types of creatures nematodes love to feast on.

FUN FACT

In the Dark

Worms love the dark and dislike light. Because sunlight can quickly dehydrate worms, most will burrow deep into the vermicompost bedding to avoid light. For this reason, your vermicomposting bin should be covered but not entirely sealed—worms need oxygen to create compost.

POOR COMPOST QUALITY

If you notice that the compost produced by your worm farm is of poor quality, it may be due to one of the following issues:

Improper Balance of Materials

It's vital to the success of your farm to maintain a balance of carbon-rich materials, such as cardboard, and nitrogen-rich materials, such as food scraps, in your worm farm. This is one of the most important considerations for maintaining a healthy population of worms. An imbalance of these materials can lead to a slower composting process and poor-quality compost.

Since your vermicompost farm functions like a small ecosystem, the components matter and influence each other greatly. As mentioned in previous chapters, a surplus of nitrogen-rich organic matter is one of the primary ways your farm's environment becomes imbalanced. Too much nitrogen leads to excess moisture and heat, attracts a variety of pests, and creates an overall anaerobic environment. This is when you can start seeing your worm farm resembling a hot pile of compost rather than a vermicomposting project. Thankfully, the solution is very simple: just add more carbon-rich bedding.

Maintaining a proper carbon-to-nitrogen and PH balance in your bin will encourage your worms to reproduce and supply them with a safe and comfortable environment in which they will continue to thrive.

Insufficient Moisture

A lack of moisture is not nearly as common a problem as an excess of it, but it's worth mentioning. If you've neglected to feed your worms for a while but continued to provide them with dry types of bedding, you will end up with the opposite problem—a worm farm that is too dry.

Your worms need just the right amount of moisture. Not excessive moisture, but enough that their bodies don't dry out. Worms need a moisture level of around 70% to flourish and produce high-quality compost. If the bedding in your worm farm is too dry, the composting process may slow down or stop altogether. To maintain the proper moisture level, mist the bedding with water as needed. So long as your bedding has the overall feel of a well-wrung-out sponge, your worms will be able to draw plenty of moisture from it.

Other ways to add moisture are to spray your bedding with a bit of coffee, feed high-moisture fruit such as watermelon, or put your fruit and veggie scraps through a blender before adding them to your bin. Liquifying your scraps will break them down and make them available to your worms faster while adding moisture to the bin.

A lack of moisture is more common in tropical locations, as the heat tends to dry out your bin quickly. In this case, keep your spray bottle handy and spritz the bedding as needed. A moisture meter can help if you are not comfortable getting your hands involved. Otherwise, use the squeeze test to keep the moisture levels in check. Ideally, your bedding will have enough moisture that only a few drops will come out when you squeeze a handful of it.

Overcrowding

It's important to maintain a balance of worms and organic materials in your worm farm. Just as an excess of organic material and food and insufficient worms can cause a plethora of problems, the opposite dynamic will also result in an unhappy worm farm.

If you have too many worms in your farm, they may not have enough space or food to produce high-quality compost. Most of the time, your worm population will slow down reproduction once space becomes limited, but there may be some adjustment pains while they figure it out. As new baby worms hatch, your adult worms may feel crowded and attempt to crawl up the sides of your container in search of more spacious lodgings.

It's easy to know when your vermicompost bin is becoming overcrowded because there will be visible congestion. This is the perfect time to gather up the hopeful escapees and start a new bin or lure a friend into the benefits of vermicomposting. Starting another bin will allow you to recycle an even greater amount of compost and yield more castings for your garden.

Poor Ventilation or Excessive Moisture

Worms need access to oxygen to thrive and produce compost. If your worm farm has poor ventilation, the worms may not have enough oxygen, leading to a slower composting process and poor-quality compost. In some types of containers, such as window-type systems or worm farms that live in breathable bags, this is not as much of a problem since excess moisture can drain away or disperse.

In your typical plastic container, excess moisture can be a real problem and is greatly exacerbated by overfeeding or feeding an excess of fresh fruits that contain a lot of moisture. These happen to be the same foods your worms love, such as melons, watermelons, or pumpkins. This is why when you feed this type of organic matter, you must take care to balance it with dry forms of bedding or feed it in small amounts.

Wynkoop shares her own experience, stating, "The biggest issue we deal with is mite outbreaks. This happens when the moisture content of the bin gets too high. We will usually remove any food sources and then add dry shredded paper on top to dry out the surface of the bin. I will keep doing this until the environment is too dry for the mites to sustain and they will eventually die off. Do not feed the worms during this process. They can eat the existing bedding and will survive without food for several months as long as they have moisture."

Excess moisture and poor ventilation can be solved by a combination of ensuring your worm farm has proper ventilation and taking care to keep the moisture level of the food and bedding in proper proportions.

TIPS FOR TROUBLESHOOTING
AND RESOLVING COMMON PROBLEMS

The most important thing to know about troubleshooting and resolving the most common problems is that paying attention to the daily activities of your worm farm is easily the most important factor in achieving success. **Scott Wilson, creator of the Gardener Scott YouTube channel,** explains his simple process: "I use my own senses to monitor bin conditions. If I suspect it's too wet, I stick my hand into the bin. I put my face close to the bin when I open it to detect any bad smells. I listen and look for active worm activity after opening the bin. I keep my bins in an indoor location with consistent temperatures. Paying attention to the worm environment doesn't require anything mechanical."

Of course, not all of us can keep our bins indoors for maximum temperature control, but there are gadgets and tools that can help us monitor the temperature of our outdoor farm. Read through this quick list of tips and keep them in mind as you continue learning about vermiculture.

- Purchase a moisture meter and temperature gauge: They will go a long way toward troubleshooting the most common problems.
- Keep an eye on your farm: Make it part of your daily routine to pass by your worm farm, so you can spot any visible problems.
- Use your nose: A rotten, funky smell points to moisture and heat excess. If your farm is in close proximity to your house or in a location you frequently pass by, you'll be more likely to notice the smell right away.
- Don't worry too much about getting rid of pests in the beginning: Focus on keeping your vermicompost containers healthy and balanced. This will prevent the majority of pest problems before they even start.

Frequently Asked Questions About Common Problems

How do I know if my worm farm is not draining properly?

Your farm will be soggy, too moist, and become anaerobic. Make sure that whatever type of bin you have selected has a good filter that allows moisture to seep through and drain well. If moisture cannot drain from your container, your farm will essentially putrefy, causing your worms to try to escape or even drown.

Why are my worms not eating the manure I've added to their bins?

They won't eat manure until after they eat the food scraps you've fed them. If you have added manure to your worm farm, be patient and allow your worm colony to establish. After the first couple of months, and after they have consumed the initial food and bedding, they will move on to the manure. The finished product will be a container full of rich vermicompost that you can use in your gardens, and well worth the wait.

How do I get rid of black soldier flies?

Black soldier flies are definitely one of the least desirable developments in a worm bin. They are not pleasant to look at and very hard to get rid of once they spread throughout your bin. In fact, it's so difficult that a complete overhaul and fresh start are often necessary.

In this case, prevention is the most successful tactic and usually comes in the form of using a fly screen to cover the top of your worm farm. You can make this screen yourself and pin it to a wooden frame that goes around the edges of your container. Ideally, you would also have a lid on top of the screen to prevent the next problem—rodents.

How can I prevent rodents from getting into my bin?

You won't run into this problem if you are vermicomposting indoors, but if you are on a farm, it's a different matter. The easiest solution is to have a well-fitting lid that covers your container so rodents cannot remove or push it off. If your farm doesn't have a lid, the alternative is building a wire mesh that acts as a lid, which rodents will not be able to get through.

Should I bury my food scraps?

Definitely not. Although this advice is commonly handed out on the internet and passed around by beginner vermicomposters, it's a good way to encourage the development of a variety of problems in your bins.

The only time I would recommend burying food scraps is as part of harvesting or cleaning your bins. For example, if you want to move your worm population to a fresh bin or relocate it temporarily while you refresh your main bin, you can place food scraps and bedding in onion bags, bury these bags, and allow them to collect your worms so you can easily move them where you want.

Why am I seeing ants or fire ants in my worm bin?

If your worm farm is outdoors and you normally have a bounty of ants in the area you've located your bin, they may migrate into your bin. However, this doesn't normally happen unless the ecosystem of your bin has become too dry or too acidic. Ants have different environmental requirements than worms and will not be attracted to your bin if it is in its naturally balanced state. Increase moisture levels and neutralize the acidic conditions by adding dry eggshells, and the ants should move out promptly.

6

LOCATION-
SPECIFIC
CHALLENGES

Vermicomposting is a versatile and environmentally friendly method of composting, but it's important to consider the specific challenges and conditions of your location when setting up a worm farm. In this chapter, we'll cover some common location-specific challenges of vermicomposting and offer tips on how to overcome them.

Vermicomposting in Cold Climates

If you live in a cold climate, you may face some challenges when it comes to vermicomposting. The first thing to remember is that if you control the temperature inside the container, the temperature outside won't have as much of an impact on your worms. It's the temperature inside your bin that really matters, and there are a variety of ways to affect it. With that in mind, here are some simple considerations and tips to help you succeed.

CHOOSE A LOCATION THAT STAYS WARM

Worms prefer a temperature range of 55 to 77°F, with the African nightcrawler being able to tolerate temperatures of up to 95°F, although not happily. Thus, it's important to choose a location for your worm farm that stays within an ideal range. It will make heat and moisture management much easier in the long run, with less work on your part.

If you can move your worm bin, it will be much easier to control the temperature year-round, regardless of the climate you live in. Thus, in a temperate climate, the best choice for a worm farm container could be a bin-type or stackable bin system. This will give you the mobility necessary to change locations as temperatures fluctuate throughout the year. With a mobile worm farm, you can even keep it in your kitchen if space is not a constraint.

If a wheeled container is not possible, a basement or unheated garage can be a good option, as they tend to be more protected from the elements. A barn or shed will also work, provided that you test the temperature in your container and use some type of insulation or bedding to keep it within a reasonable range.

When considering whether your worm farm will be located indoors or outdoors, also give thought to the size of the container you use. In cold climates, it will prove easier to maintain even temperatures in a larger container than in a smaller one. While a small container will lose heat very quickly, a large one will retain heat, and your worms will be able to burrow into deeper layers of soil to keep themselves warm.

INSULATE THE CONTAINER

There are various techniques to insulate your vermicompost to help maintain a consistent temperature. Generally, the two options are to either insulate the outside of your container or to insulate the inside. Best results will likely come from using both tactics, depending on just how low temperatures drop in your location.

To insulate around the outside of your container, you can use a variety of materials to create a buffer against the cold: bales of straw, Styrofoam, and blue insulation foam work best. Other materials, such as bubble wrap, cardboard, tarp, or leaves, can also be used but may not work as well as the heavier insulation options.

Another choice, if you can only keep your container outdoors, is to bury it completely or at least partially so that when temperatures drop, your container will be insulated by the soil around it, sheltering your worms from the harshest effects of cold temperatures. **David Trodd, The Weedy Gardener**, knows that even if "the worms die during a cold winter, the eggs will survive, and the population will increase once the temp rises again in the spring."

There are also many reports of old freezers being used as vermicompost containers; they come with built-in insulation and may serve as a viable option in cold climates. Just be sure to make good drainage and aeration holes if you choose this option. Since a freezer is sealed, it will be more difficult to circulate air throughout, which can result in an excess of moisture.

> *In a cold climate, it is important to make sure that the worms don't freeze to death, so I recommend keeping them surrounded by bales of hay or bringing them indoors during the winter. If the worms do die during a cold winter, don't worry—the eggs will survive and the population will increase once the temperature rises again in the spring."*
>
> **David Trood**
> The Weedy Garden

ADD ADDITIONAL BEDDING

To help maintain the proper moisture levels and provide insulation, you can add additional bedding to the inside of your container. Shredded newspaper or coconut coir are some of the best choices. You can also use layers of presoaked newspaper or cardboard to increase the temperature of your container, which will help keep the worms warm and active during the colder months. This works by creating a thermal barrier to slow heat from escaping out of your bin. Simply layer several sheets of moistened newspaper on the very top of your bin's contents.

During the coldest months, you can fill your container right up to the top with fluffy bedding such as shredded newspaper or leaves to retain heat. Using bedding with higher nitrogen content, such as brown leaves, mulch, and aged manure, will help to increase the temperature in your bin, as will covering the top of the container with a worm blanket. A do-it-yourself blanket can be as simple as using burlap sacks normally used for coffee beans. Just lay it across the very top of your worm bin to keep your worms nice and snug.

AVOID OVERFEEDING

Overfeeding can lead to an excess of uneaten food and a buildup of harmful bacteria. In cold temperatures, the worms' metabolism slows down, so they will not be able to process large amounts of food. It's important to feed the worms an amount that is appropriate for their size and activity level.

If your farm is outside in the cold winter months, your worms' activity may slow down so much that it will be better to do less rather than more. This means minimal feedings and minimal disruptions while your worms are essentially hibernating through the lowest temperatures.

Vermicomposting in Hot Climates

Alternatively, if you live in a hot climate, you may face some heat-related challenges when it comes to vermicomposting. Here are some tips to help you succeed and see your worms through the hottest months of the year.

CHOOSE A LOCATION WITH SHADE AND VENTILATION

As worms prefer a temperature range of 55 to 77°F, it's important to choose a location for your worm farm that allows your bin to remain within this range. A shaded location can help keep the container cool and prevent the worms from overheating.

If you can combine shade with ventilation, it would be ideal. In our tropical climate, we've chosen a location that is among many trees, including some evergreens that provide shade year-round. An area surrounded by trees will keep temperatures lower not only by blocking the sun but through the trees' natural process of transpiration, which gives off a cooling mist and keeps the area around the trees at a lower overall temperature.

For those living in hotter climates, full shade and ample bin size is critical. I like to place bins on the north side of a house, under dense tree canopy, or, if you're working in an open, sunny space, beneath some dense shade cloth so they get full shade protection."

Elise Pickett
The Urban Gardener

Placing your container on the north side of your house will provide additional help and safeguard it against the harshest sun of the day. A covered and breezy porch would work as well, and even an awning would give some respite from the sun.

In addition to choosing a shady spot, you can paint the top of your containers white or very light colors to make sure they are not absorbing additional heat.

ADD ADDITIONAL BEDDING

To help maintain the proper moisture levels and provide insulation, you can add additional bedding to the container. Shredded newspaper and burlap, also known as hessian or coconut coir, are some of the best choices for bedding material in hot climates.

The best type of bedding to use in high temperatures is absorbent material that both holds the moisture and allows for aeration. Try to stay away from materials that absorb a lot of moisture but remain soggy and heavy with water, as these are more likely to rot and cause further problems.

MIST THE BEDDING

To help keep the worms cool and prevent the bedding from drying out, you can mist the bedding with water. Be sure not to oversaturate the bedding, as this can create an excess of standing water and create a breeding ground for pests.

Using a spray bottle with a fine mist will suffice. Spray a little water at a time, and then observe your bin over the next day to see how long it takes for the moisture to evaporate. As you get better at understanding the effects of your actions on your worm bin, keeping the environment in balance will get easier.

AVOID OVERFEEDING

Overfeeding can lead to an excess of uneaten food and a buildup of harmful bacteria. Just as in cold temperatures, the worms' metabolism slows down, so they will be unable to process large amounts of food. As in cold temperatures, make sure you only feed the worms an amount that is appropriate for their activity levels, even if this means very little food.

When you do feed the worms during the hottest months, ensure that you leave large sections of your bin free of food so that your worms have sufficient space to stay cool away from decomposing food.

Lightning Speed

Vermicomposting can accelerate the decomposition process by two to five times and may produce higher-quality soil than regular hot composting. For example, you can expect your worms to produce a measurable amount of vermicompost every six to eight weeks, while traditional composting can take around six months or more to mature.

Vermicomposting in small or urban spaces

If you live in a small urban space, you may face some challenges when it comes to vermicomposting. The good news is that most of these problems will be specifically related to space rather than to temperatures, and in our humble opinion, space-constraint problems are much easier to resolve.

In an apartment, your worm farm could happily live in the kitchen, perhaps under the counter if it's small enough, or in an unused area of the kitchen if you are using a vertical system. You may also want to consider a balcony if this is an option. In this way, your vermicompost farm would be out of the way for most of the year, only having to be brought into your living space during extreme temperatures.

HELPFUL TIP

Urban Composting

Because vermicomposting is odorless and compact, it is one of the most accessible indoor composting options for apartment dwellers. In addition, indoor worm farms can be highly compact, perfect for people with space at a premium. Today, many attractive and compact vermi-composting bins are commercially available, making this method of kitchen waste recycling an accessible and appealing option.

CHOOSE A SUITABLE CONTAINER

Choosing a container that is appropriate for your needs will help you make the most of your small space. A stackable container or a vertical worm tower can be a good option. A vertical or tray-based system allows you to get the most castings in a small space and will also be easier to clean and maintain.

Most vertical systems allow you to feed your worms in the topmost container while being able to harvest the container at the bottom of the system. Harvesting castings will naturally be easier in this type of system because the bottom tray you harvest will have the fewest worms. In an indoor setting, this method saves you a lot of messy sorting and cleaning. Your family will be more likely to enjoy your vermicomposting project indoors if it comes with the least amount of mess to clean up.

USE A FLOW-THROUGH SYSTEM

A flow-through system can be a good option for small spaces, as it allows you to continuously add organic matter while expending the least amount of effort in managing your vermicompost farm.

There are a variety of flow-through systems available for purchase that will do well in small, compact spaces. Some of the most popular ones are the Worm Bucket, the Essential Living Worm Composter, the Worm Factory 360, and several other models that all function in similar ways and are designed to save space and minimize your work. These

I use a flow-through system called the Urban Worm Bag. I chose that product because it is easy to use and allows me to keep my worm farm in my house year-round. I also like that it makes it easy for me to keep adding material to the top and extract finished castings from the bottom."

Mark Townsend
Cousins Compost LLC

can all be purchased online and sometimes in large gardening centers.

As with most products that are meant to minimize labor, flow-through systems come with a higher up-front cost. It's the price of convenience, but in small spaces, these products can make a significant difference in the disruption and mess your home will endure.

TIPS FOR VERMICOMPOSTING IN DIFFERENT LOCATIONS

- Choose a container or system that works well in your environment, whether this means you are in a hot climate, a cold climate, or a small apartment in a busy city.
- In small spaces, set up your worm farm in a way that keeps cleaning and messes to a minimum, as it will go a long way toward making you feel amiable towards your worm friends.
- Take advantage of the natural features of your environment: use shade and the cooling effect of trees to keep your worms happy.

7

EXPANDING YOUR VERMI- COMPOSTING OPERATION

Once you've gained some experience with vermicomposting and are producing high-quality compost, you may want to consider expanding your operation. In this chapter, we'll cover some options for expanding your vermicomposting farm and how to scale up successfully.

Increasing the size of your worm farm

One way to expand your vermicomposting operation is to increase the size of your worm farm. This can be especially useful if you have a high volume of organic waste to compost or if you want to produce more compost for use in your garden or landscape.

The basic steps for expanding the size of your vermicompost farm are as follows:

- Choose a larger container: You'll need a larger container to accommodate the additional worms and organic waste. Consider using a bin system or a flow-through system, depending on your needs and resources.
- Increase the number of dwellers in your vermicompost farm: Either purchase more worms or gradually raise the number of worms in your current farm as you increase feedings and the space available.
- Keep a close eye on your vermicompost operation as you are expanding it to make sure neither the quality of castings nor the health of your worms suffers as you scale up.

As you expand the size of your worm population and the amount of organic waste you provide, give thought to the amount of work a larger farm will require. **Kelly Hammel, who runs the Worm Hippie Worm Farm**, suggests that you ensure you "have the time, space, and automated equipment to scale up the operation. Worm farming is labor intensive, so making some jobs easier and faster is a must when having more work to do."

Because worm populations can grow quickly, starting with a small number of worms allows time to learn how and what to feed worms while maintaining a healthy environment. After harvesting the first batch of castings, the worms can then be split between two bins for multiplying. Two bins will introduce new requirements for food, time, and attention and can help a vermicomposter determine how big an operation they wish to maintain."

Scott Wilson
Gardener Scott

Choose a system that is as self-sustaining as possible, whether due to its larger surface area or because it is a continuous-flow system that requires less hands-on maintenance and operation. A flow-through system that lets you harvest from the bottom and feed from the top will require the least amount of heavy labor and be easier to operate on a large scale.

Of course, you can also use simple concrete or block containers, which is a very popular and affordable option in many parts of the world. Even though these systems are sizable, they can be harvested by removing the top layer of worms into another bin and then using a tool like a shovel to remove the vermicompost from underneath. It requires a little more manual work but also has a much lower up-front cost. In the center photo below, you can see two of our concrete bins being set up.

HELPFUL TIP

Bedding

Bedding is a crucial component of any worm farm and can be purchased or created from common household materials. When choosing bedding for your worms, it's critical to provide bedding free of harmful chemicals, odorless, pH neutral, and capable of retaining moisture. Popular bedding choices include shredded cardboard, wood chips, straw, or shredded newspaper. Commercial vermicompost options are also available.

POPULAR OPTIONS FOR LARGE-SCALE VERMICOMPOST CONTAINERS

Type of Container	Pros	Cons
Continuous-flow vermicomposters 	• Incredible efficiency! • Easy temperature regulation: Many CFT systems come with built-in controls or insulation. • Very little labor involved in maintaining and harvesting • Some CFT systems are modular, so they are very easy to expand.	• More complex operation, which can be challenging for beginners • High initial cost • Produces very clean castings
Large concrete or wooden containers 	• Can be built to any custom size • Concrete provides good insulation in both hot and cold weather. • Easier to maintain than pits or trenches since you can build these containers to be thinner • Generally inexpensive	• Cost of materials and cost of labor to build • Wood containers will eventually rot if not treated properly. • Can flood if not deep enough and protected from rain
Pits or trenches (often made in hog pits, but can be built as well) 	• Easy to make, no complicated construction needed • Naturally provide insulation against hot and cold weather • Most economical option • Can be labor intensive to set up if no existing trenches	• Difficult to reach the entire surface, a lot of bending required • Can be flooded by heavy rains since they are low to the ground

ADD MORE WORMS

To handle the increased volume of organic waste, you'll need to add more worms to your worm farm. Red worms or night-crawlers are both good options, but take your climate into consideration to decide on which type of nightcrawler to use.

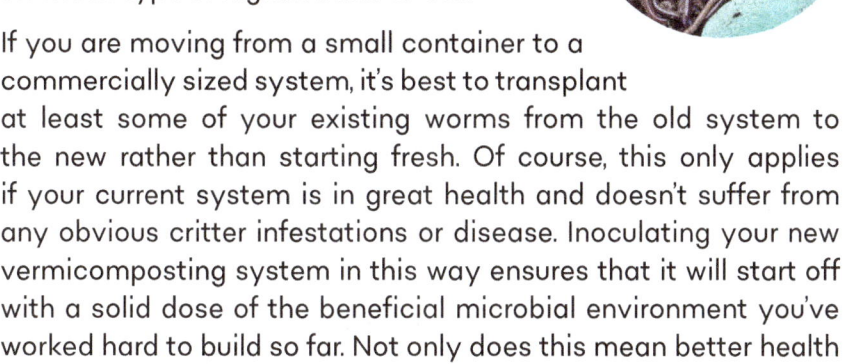

If you are moving from a small container to a commercially sized system, it's best to transplant at least some of your existing worms from the old system to the new rather than starting fresh. Of course, this only applies if your current system is in great health and doesn't suffer from any obvious critter infestations or disease. Inoculating your new vermicomposting system in this way ensures that it will start off with a solid dose of the beneficial microbial environment you've worked hard to build so far. Not only does this mean better health for your new system, but also quicker expansion.

INCREASE THE AMOUNT OF ORGANIC WASTE

As you add more worms, you'll also need to increase the amount of organic waste you add to your worm farm. Be sure to maintain a balance of carbon-rich materials, such as cardboard, and nitrogen-rich materials, such as food scraps.

Once you are actively growing your worm farm for commercial use, you will have to either increase the amount of organic waste you produce or get a little creative in sourcing it. The first option is easier if you have a large family and eat a lot of fresh food. Just remember that whatever you feed your worms in organic matter must be balanced out with higher-carbon materials like cardboard, newspaper, or woodchips.

To increase the amount of organic food you have for your worms, you can involve your friends and neighbors and ask them to save their organic waste for your vermicompost farm as well. Coffee shops and restaurants are another great source of free worm food since they often have a lot of food scraps that will end up in landfills anyway.

MONITOR AND ADJUST AS NEEDED

As you scale up your worm farm, it's important to monitor the health of the worms and the quality of the compost. Make any necessary adjustments to maintain a healthy environment for the worms and optimize the composting process.

It will be even more important to take basic measurements of your worm farm's temperature and moisture levels as your project grows. A temperature gauge, one that uses a wired probe that can be inserted in your bin, would be best. This way, you are measuring the temperature of the compost rather than the air temperature.

A moisture meter would be helpful as well. This can easily be moved around your worm bin to test the moisture level in different spots or even in different bins. The more worms you have to care for, the easier your work will get if you put some basic monitoring systems in place.

FUN FACT

A Plethora of Hearts

Red wigglers have five hearts. Though much less complex than a mammalian heart, these earthworm heart-like structures deliver oxygenated blood throughout the worm's body.

SECURING ADDITIONAL MATERIALS FOR YOUR GROWING WORM FARM

Type of Organic Materials	Where to Procure
Greens (30%)	
Kitchen scraps (only fruits, veggies, coffee grounds, eggshells)	Your own kitchen, restaurants, cafés, soup kitchens, grocery stores, food or farmer's markets
Aged animal manure	Nearby farmers, horse stables, horse racing tracks, horse riding farms, agricultural centers or farms, online classified ads
Grass clippings	Your own yard, neighbors, landscaping companies, farmers, local transfer stations
Browns (70%)	
Dry garden bedding (brown leaves, straw, small sticks from branches, straw, or hay)	Your own property, neighbors, tree cutting companies, landscaping companies, farmers, local transfer stations
Paper and cardboard (no ink)	Printing shops, big-box stores, pharmacies, bookstores, recycling centers, liquor stores, arts and crafts stores
Egg cartons	Egg farmers, recycling facilities
Woodchips, sawdust	Local workshops or woodworkers, local schools, tree cutting companies, electric or utility companies

Using vermicompost on a larger scale

If you want to use your vermicompost on a larger scale, there are a few options to consider.

USE IT TO FERTILIZE A LARGER AREA

If you have a larger garden or landscape, you can use your vermicompost to fertilize a larger area. Simply mix it into the top few inches of soil around your plants. Use it as a rich amendment to start new seeds by adding it to your seed-starting soil or around the new plantings. If you are planting directly outdoors, add it directly to the holes you make for your seeds.

When producing a large amount of vermicompost and using it on a large area, you will need to find a good way to store it. Make sure it stays somewhat moist but not completely wet. It's also important that whatever type of container you store it in, whether a plastic bin, bucket, or sack, has airflow. Make holes if you need to.

SELL OR DONATE YOUR VERMICOMPOST

If you're producing more vermicompost than you can use, you can sell or donate it to others. This can involve creating a website or networking with local farmers and gardeners. While you can create a simple website yourself with the help of YouTube and online instructions, a more complex, business-quality website will require outsourced help, particularly if marketing is required.

The good news is that selling vermicompost is more lucrative than selling compost! In The Worm Farmer's Handbook, Rhonda Sherman points out that while compost sells for up to $35 per cubic yard, vermicompost sells for a whopping $200 to $1800 per cubic yard (Sherman, 2018). That's an impressive difference, and it's reflective of the incredible benefits of vermicompost.

Donating your excess vermicompost won't be difficult. There are many institutions and individuals who would love to receive a free and all-natural boost to their growing efforts. Start close to home in your local neighborhood, and see if a nearby school is running a gardening program for students. Vermicompost could be just the help they need to increase their yields. Or perhaps there is a local farm nearby that could benefit from this top-of-the-line natural fertilizer. Vermicompost is not inexpensive to buy, and local farmers are surely trying to keep operating costs low. Other venues that might benefit from your vermicompost could be colleges or universities, community gardens, hospitals, or individual households.

STEPS TO GROW YOUR VERMICOMPOST BUSINESS

Increase the size of your container

Increase the number of worms

Increase the amount of organic waste

Put delivery and packaging systems in place

Market your product and connect with potential buyers

Sell or donate your vermicompost

SET UP A VERMICOMPOSTING BUSINESS

This can involve collecting organic waste from commercial sources, such as restaurants or grocery stores, and processing it into compost using worms. To set up a vermicomposting business, consider the following steps:

Research the market

Determine the demand for vermicompost in your area and assess the competition. This will help you understand the potential for your business and how to differentiate yourself from other vermicomposting businesses.

Are there large potential buyers in your local area? Some of the biggest consumers of vermicomposting materials are home gardening centers, farms, landscaping companies, and golf courses. Farmers with especially high-value crops, and particularly organic farmers, could be an especially receptive market for your vermicompost, as they will be most interested in maximizing their yields without the use of chemical fertilizers.

If you live in a remote area without any of these types of businesses, research the costs involved in shipping your vermicompost to different areas and to varying distances. Shipping costs will make up a substantial portion of your operational costs and must be factored into your overall expenses.

Are there other commercial vermicomposters in your local area? Are they also selling to a local market? If there are competitors, but they are only selling remotely, then it won't matter as much. The remote market is large, as large as your ability to support shipping costs, and can offer you at least a multi-state or even national market.

Acquire the necessary equipment and materials

You'll need a large-scale worm farm, as well as a transportation system for collecting and delivering organic waste.

There will be some higher initial costs as you are first setting up your commercial farm. These will include the cost of constructing or purchasing the vermicompost beds or containers, any machinery you will use to sort the vermicompost, and perhaps the building of a shed or other structure that will house your farm.

While a small vermicompost bin won't take up a lot of space, and the sorting and maintenance may not require a dedicated area, a commercial system needs more up-front consideration. You'll not only need space for your worm bins but also for storing the organic matter you collect, whether this is only food scraps or mixed with manure and other types. If you are collecting organic matter from outside sources, you will need an area where you can sort it, cut it, or chop it into smaller bits if necessary. If you will be using a sorter to harvest your vermicompost, that machine will also need its own space. Consider all components that make up the processing system of your vermicomposting operation, including any storage space you will use for the finished vermicompost as it waits for sale, donation, or shipment.

You will also need to secure transportation for the collection of organic waste, animal manure, and other materials you will use in your bins. Additionally, transportation will have to be arranged for the finished castings.

Establish partnerships

Consider partnering with local businesses, such as restaurants and grocery stores, to collect their organic waste and turn it into compost.

Many businesses are learning how to go green and get involved in green initiatives. Speak to restaurants and grocery stores in your city and ask them if they would like to reduce their waste and become more eco-conscious. Arrange to have their organic waste picked up on a regular basis and turned into quality vermicompost that will go on to help their own community grow tastier and healthier food.

Market and sell your compost

Once you're producing high-quality compost, you'll need to market and sell it to potential customers. This can involve creating a website, networking with local farmers and gardeners, and participating in local events. **Caleb Johnson of Permies Worms** points out, "Marketing is key! You want to make sure you are educating customers. Even though worms have been around forever, people still don't understand their true powers."

Marketing in a highly digital world has changed form over the years, and social media content has a much larger importance than in previous years. In addition to creating a website, which every business should have, you can create additional content to distinguish yourself from other local vermicompost businesses. You can start a YouTube channel with instructional videos or videos detailing your day-to-day operations. Explainer videos on how others can set up their own vermicompost farms, testimonials from clients who use your vermicompost successfully, or customer reviews all go a long way in building trust with your potential clients.

Participating in local farmer's markets and gardening events where you can offer samples of your castings and let people in on the benefits of your product is also a fun and effective way to get involved in your local community and build your reputation. There are many other types of community events you can get involved in, such as school events, street parties, or volunteer events. You can even plan your own event and invite people in your local community to a big cookout and a visit to your vermicompost farm.

Tips for expanding a vermicompost operation

- First and foremost, be sure you are producing quality castings before you prepare to expand your worm operation.
- Be mindful of what you feed your worms, as input of organic materials will determine the quality of the vermicompost you will harvest.
- Put a system in place that monitors the health of your farm: this can include daily "sniff" checks, temperature gauges, and moisture meters.
- Keep your worm beds in the ideal temperature range so that they will reproduce faster.
- Don't be afraid to get involved in your community to make business contacts and expand your commercial vermicompost operation faster and further.

8

VERMICOM-POSTING WITH CHILDREN

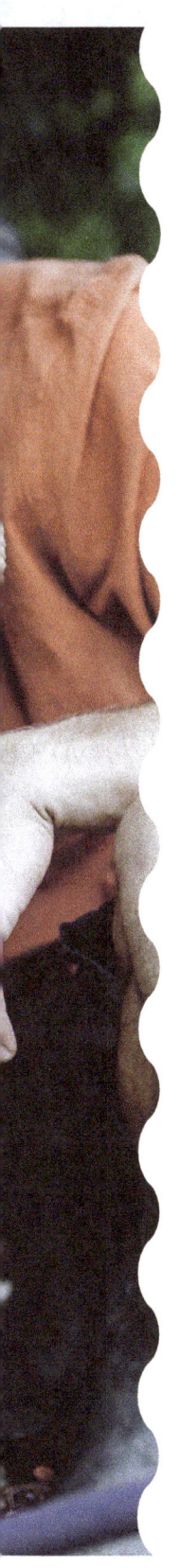

Involving children in worm farming gives them an opportunity to rehearse their growing need for autonomy and independence while learning how to care for living and nonliving things. It's a wonderful opportunity to collaborate with others, learn about our individual impact on the environment, and explore the natural world.

BENEFITS OF VERMICOMPOSTING WITH CHILDREN

Encourages independence, autonomy, and personal responsability

Offers a diverse range of educationational and hands-on activities

Guides children into an appreciation for living things and the natural environment

Teaches science concepts and opens opportunity for scientific experiments

Educational Activities

Kids love worms. They're wiggly, they're fun to play with, and they are easy and resilient enough to care for that even the youngest learner can manage it without doing harm.

EDUCATE THROUGH VERMICOMPOSTING

A variety of educational activities can be incorporated into vermicomposting with children, such as worm identification, composting experiments, journaling, and building a miniature or reduced version of a vermicompost bin.

As there are several thousand different species of worms, worm identification can be a fun activity for kids. You can use drawings, diagrams, or videos to talk about different types of worms. Locate some common earthworms and compare their physical features to your vermicompost worms. Talk about the difference in the behavioral patterns and physiological needs of regular worms versus the red wiggler or nightcrawler worm.

Composting experiments can take many forms. Curious about what types of foods your worms like more? Place two different types on opposite ends of your bin and see which one they gravitate to first. Want to find out how fast your vermicompost worms eat? Add different types of organic matter to your bin and monitor how long it takes your worms to eat it.

To incorporate gardening experiments in your activities, test the effects of vermicompost on growing plants or vegetables. Add your castings to some and leave others without. Observe what effects vermicompost has on your garden. Do your plants and veggies grow faster? Healthier? Ask your young vermicompost pupils to record their findings or journal their observations.

Build a small version of a vermicompost bin with your youngest kids. You can use a small plastic container, a plastic-lined shoe box, or a Styrofoam cooler, something easily maneuverable by little hands. Set up your miniature vermicompost container the same way you would set up a regular worm farm. Show your littlest ones how to gently handle the worms and add little bits of food for their worm friends.

For older pupils, show them how to track the progress of your worm farm in detail: record types of bedding and organic matter used, moisture and temperature levels, worm behavior and response to the environment, and the effects of vermicompost on your garden.

Vermicomposting is a great way to acknowledge the role that soil organisms play in successful gardening. Understanding how worms take scraps and turn them into a wonderful soil amendment makes it easy to recognize the role that native earthworms and other soil organisms play in one's garden."

Scott Wilson
Gardener Scott

VERMICOMPOSTING IN THE CLASSROOM OR HOMESCHOOLING CURRICULUM

Whether your children are in school or learning at home, important lessons can be learned through vermicomposting, such as gardening skills and concepts regarding personal responsibility, self-stewardship, and self-sustainability.

Teaching gardening to children awakens them to their own power as stewards of this world and of their own lives. It teaches both practical, objective skills required to grow food and broader concepts about the power we all have to affect what we eat, the process of producing more of our own food, and what it means for us in our communities.

Basic gardening lessons can start with the benefits of vermicomposting on the food and vegetables in our own gardens and the value of using natural and nontoxic ways of helping plants grow. Lessons can include plant selection, planting, weeding, mulching, and even cooking.

Getting children involved in your vermicomposting farm can be as easy as encouraging them to connect to the project on their own terms. Are they busybodies who like to get their hands dirty? Lean heavily on the hands-on gardening aspects and teach them the ins and outs of how things grow.

Do you have older children who like to read and write? Talk to them about the higher concepts of vermicomposting and ask them to brainstorm their own ideas about how vermicomposting contributes to a self-sustaining model of agriculture. Together, talk about what it means to participate in building agricultural models that encourage personal responsibility, the value of organic farming practices, and how this benefits our society.

TEACH SCIENCE CONCEPTS THROUGH VERMICOMPOSTING

Vermicomposting is a fantastic, engaging way to teach science concepts, such as life cycles, nutrient cycling, decomposition, ecosystem diversity, and diffusion.

Through the observation of your worm farm, you can teach young learners about the life cycle of worms and other creatures present in your worm bin, as well as nutrient life cycles and the process by which worms can transform our garbage into high-quality natural fertilizer. Explain that although different organisms and animals have different life cycles, they all share the commonality of birth, growth, reproduction stage, and death. Try to identify worms in their different stages of life, from eggs to juvenile stage to full maturation and reproduction.

Perhaps the most exciting role of vermicompost worms is their ability to affect nutrient cycling and recycle discarded organic matter into what vermicompost enthusiasts refer to as the black gold of our gardens. This repurposing of nature's leftovers happens through a microbial process that is essentially the digestion of organic matter and excretion of the casting.

Constructing a worm bin from scratch is the easiest way for kids to understand how vermicompost is made. Setting up the bin together, getting the organic materials and bedding ready, feeding the worms, and observing the decomposition process will allow kids to study the entire system from start to finish. Ask them to pay particular attention to the other creatures that appear in your bin. Research their role in the vermicompost process together, and discuss how they help or hinder the worms' work.

> *Worms are amazing, simple creatures, and are a joy to work with. Every day is a learning experience when worm farming, and worms can help you teach about biology, soil health, composting, upcycling, and many other important lessons surrounding sustainability and the environment."*
>
> **Scott Wilson**
> Gardener Scott

Talk about the physical features of worms and explain that the hardworking creatures in your worm bin breathe through their entire bodies through a process of diffusion. Emphasize that your vermicompost worms need moisture in their habitat just as humans need moisture in our lungs because it is vital to their breathing. If your worms' bodies become dry, it interferes with their basic processes, and they can suffocate. This is why maintaining a high level of moisture in your bin is so vital to the well-being of your worms.

Safety Precautions

Worm farming is not an especially dangerous activity, so safety precautions are limited to basic things like taking care not to let children handle dangerous equipment, encouraging them to wash their hands at the end of their work, not touching rotting organic matter and manure with bare hands, and making sure they handle worms in a gentle way.

BASIC SAFETY

When small children are involved, small tools will be required. Mini shovels, spades, and rakes typically used for gardening will serve small hands well and allow children to move soil, look for worms, and move organic material in the bin. Older children can use regular tools if they are comfortable doing so.

No young children should handle heavy farming equipment used in vermicomposting, such as worm harvesters. Such equipment is best kept locked away in a shed unless being used by an adult or by older children under close supervision.

HANDLING AND DISPOSING OF MATERIALS

Children who are old enough to participate in vermicomposting will also be old enough to understand the importance of wearing gloves when sifting through organic matter. Keep either small gardening gloves or disposable plastic gloves on hand to avoid little hands being irritated by directly touching manure or food that is decomposing.

Handling the worms themselves is oftentimes one of the highlights of worm farming for kids. Their tactile little hands can hold the worms and observe how the worms feel. Are they slimy? Wet? What physical features can they detect while touching the worms? Being able to get their hands involved is an important part of the experience, as it is an important part of any activity that brings us closer to nature, and as such, should be allowed—as long as little hands are washed with soap afterward!

HANDLING WORMS

Help young kids understand the fragility of these small wiggly creatures by showing them how to handle them gently. Let children know that worms have to stay moist, so they can only be held for short periods at a time. If necessary, intermittently spray the worms with water while children are handling them or observing them as part of an activity.

Maintaining A Child-Friendly Worm Farm

Ensuring kids have a safe experience taking part in vermiculture is easy. Make your worm farm accessible and appealing to kids by following some of these tips we've learned while vermicomposting with our two young children:

- Small children will need small tools, such as mini shovels and rakes.
- Starting with a small version of a worm farm will make it easier for younger kids to participate in setting up the farm, maintaining it, and harvesting the castings.
- Triage your organic matter before making it available for your worm bin: no sharp objects, no harmful chemicals or substances.

HELPFUL TIP

Education Opportunity

Vermicomposting is a fun and educational opportunity for all ages and provides an opportunity to teach about the complex ecosystems in your backyard.

Managing the Worm Bin with Children

Managing your worm bin with children is easier than it seems. Children are naturally investigative, and their curiosity about all "adult" work makes them perfect companions for this type of activity.

TIPS

Make children feel like they are a part of this adventure, and they will be excited to participate. Kids want to feel like their opinion and their involvement matter, so make them part of the process from the beginning. Ask for their help in choosing the type of worm bin you will use. Give them exclusive responsibility for small tasks such as placing worm food in the bin or misting the bin with water on a designated schedule.

Encourage them to inspect the worm bin and its inhabitants daily so that they become familiar and comfortable with the comings and goings of all that takes place in the bin. Give them clear boundaries and simple rules for how the worms can be handled and taken care of so they feel relaxed and capable when they are working with you.

INVOLVE CHILDREN IN THE PROCESS

Involving children in the planning and design of the vermicomposting system starts with a general awareness of the different components of a worm farm.

What types of materials need to be collected for the worm bin? Encourage children to help collect leaves, scrap paper and cardboard, and other bedding materials you will use in your farm. Use the opportunity to discuss basic concepts like the decomposition cycle, the variety of materials that can be used for bedding, and their different rates of decomposition.

EASY WORM FARM ACTIVITIES FOR KIDS

- Identify worms in different life stages: adult, baby, or egg.
- Count the worms in your bin once a month and see how fast your farm is growing!
- Measure your worms and write down your findings: how long is your biggest worm?
- Record your worms' food preferences: What fruits or vegetables do they eat the fastest? How long does it take for the food to disappear completely?
- Walk around your garden and identify organic matter that can safely be repurposed and used as bedding for your worm farm.

ENCOURAGE RESPONSIBILITY

Encouraging children to take responsibility for the care and maintenance of the worm bin follows the same principles as encouraging children to follow through on any task or plan.

The most straightforward and successful way is to create a daily responsibility chart solely for your vermicompost bin, detailing the tasks that need to be completed each day. Kids can participate in making and decorating the chart with stickers, magnets, or other fun elements to show the activities they have completed.

Often, daily tasks will consist of a walk-by "sniff test," checking moisture and temperature levels, and other simple activities children can easily do independently.

SYNOPSIS

Children have an innate curiosity and affinity for living things. Encouraging their natural inquisitiveness and interest in learning will help them understand their role in the natural world. Vermicomposting can show them how they can participate in an activity that ties them into the larger community and contributes to a cleaner, healthier environment for us all.

DID YOU KNOW?

Breathing Skin

Have you ever wondered why there are so many earthworms after a rainstorm? Because worms breathe through their skin, these small invertebrates find breathing harder when rain saturates the soil, so they come to the surface for air.

9

FUN AND INTERESTING FACTS ABOUT VERMICOM-POSTING

In addition to the hard work your worms are doing in your bins, they have other useful applications: helping to clean up dangerous oil spills and reducing methane gas in the atmosphere. They also have interesting physical features that make them ideal for use in vermicomposting, including their keen sense of touch, the complex digestive systems that span their entire bodies, and their ability to eat voraciously.

History of Vermicomposting

- Vermicomposting as we know it is believed to have been introduced by a Michigan native named Mary Appelhof. Appelhof began vermicomposting with a batch of worms she bought at a bait shop, and she later went on to publish several mailers and one of the best-known books in the industry, *Worms Eat My Garbage*.

- Dr. Clive Edwards, now known as an international earthworm expert, began his research on continuous flow vermicompost systems in the 1970s. Dr. Edwards went on to write an extensive academic book, *Vermiculture Technology: Earthworms, Organic Wastes, and Environmental Management*. His book is an invaluable resource for those who want to set up successful commercial systems of vermicomposting.

- These days, vermicomposting is a global affair. Large-scale vermicomposting is being used across the world, from the United States to India, Canada, Japan, Italy, Turkey, Malaysia, the Philippines, and more.

FUN FACT

Baby Worms

Although all red wigglers possess both male and female reproductive organs, it takes two of these worms to reproduce. An adult red wiggler can lay one egg capsule containing an average of three worms per week.

Worms have superpowers

Worms have several superpowers that make them ideal for vermi-composting: in an ideal environment, they can eat their body weight in food, they have multiple stomachs, and they have an incredibly sensitive sense of touch. These traits make them very productive and ideal members of your vermicomposting farm.

THEY CAN EAT THEIR OWN WEIGHT IN FOOD EVERY DAY

Many people believe that worms can eat their own weight in food each day, and in theory, this can be true. Some use a more conservative assessment of half their weight in food per day. We can't personally tell you how much your worms will eat because we know from experience that these estimates are based on ideal scenarios. In these types of estimates, we are assumed to have perfectly balanced temperatures in the environment in our bins, an established and hearty worm population, and an ideal blend of food, bedding, and other matter.

You, as a beginner vermicomposter, will have to conduct your own experiments to determine how much your worms will eat. You'll have to remember that when we say food, we also mean bedding, manure, soil, and anything else your worms will consume. This means you cannot mix one pound of worms with one or even half a pound of organic matter each day and expect to see good results. Instead, your worms will suffocate under a hot and sticky mess of uneaten food.

Yes, red wigglers are voracious eaters. This makes them excellent at processing organic waste and turning it into compost. However, when temperatures get too hot or too cold, even red worms will slow their activity down to a crawl. Take your time and start slowly. Feed your worms small batches of organic matter at one time when you start—no more than 25% of your worms' weight in food. Then, observe. Take notes and continue with the benefit and confidence of your own experience.

THEY HAVE MULTIPLE STOMACHS

Worms have a complex digestive system that includes multiple stomachs. This allows them to break down organic material and extract the nutrients it contains. Their digestive system is made of several different parts, allowing for distinct functions that, together, enable the worm to digest different types of organic matter efficiently.

The worm swallows food through its mouth, and the food then passes through to the pharynx. This is where the digestive process begins. The pharynx releases saliva that contains mucus and proteolytic enzymes that begin to break down the food (Mlblevins, 2010).

Once food has passed through the pharynx, it continues to the esophagus, where calcium carbonate is released to destroy any acids found in food and rid the worm's body of excess calcium. After the esophagus, the food then moves to the crop, which has the same function as it does in a bird, to store food until it moves to the next part of the worm's digestive system, the gizzard.

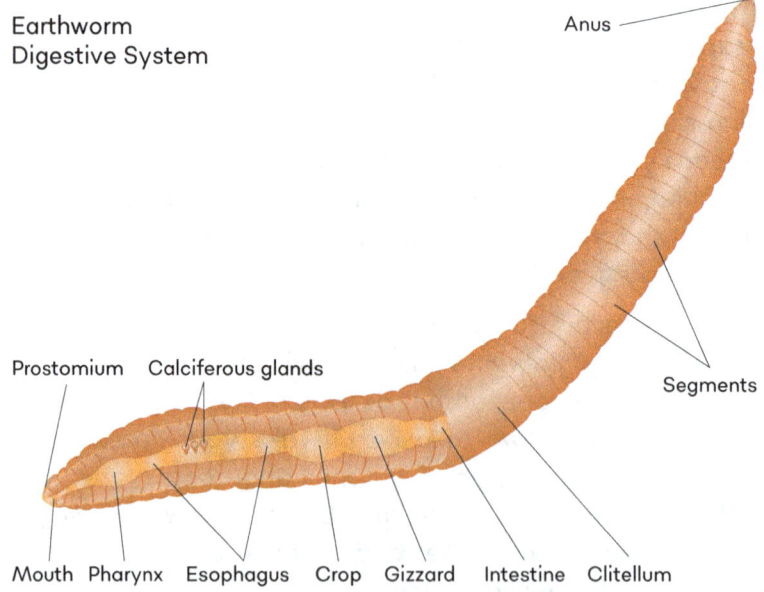

Earthworm
Digestive System

Anus

Prostomium   Calciferous glands

Segments

Mouth   Pharynx   Esophagus   Crop   Gizzard   Intestine   Clitellum

The gizzard, essentially a muscular organ that processes food, mixes it and blends it into smaller pieces so that nutrients can be absorbed and assimilated by the worm. After the gizzard, food passes into the stomach, where it is further broken down by bacteria and a variety of intestinal enzymes. The intestinal tract of the worm takes up most of the space in the worm's body. Once food has passed through this elaborate process, it exits through the anus and becomes what we know as worm cast, the rich and much sought-after vermicompost gold we use to enrich our plants and gardens.

THEY HAVE A SUPER-SENSITIVE SENSE OF TOUCH

Worms have a highly sensitive sense of touch that allows them to navigate their environment and find food. Since they lack ears and eyes, they rely on their acute sense of touch to detect vibrations. The worm's body is covered with chemoreceptors, which it uses to translate sensation into taste. These chemoreceptors are part of the worm's nervous system, used to monitor changes in temperature, chemical composition of soil, and movement in the form of vibrations. This is especially useful in a worm farm, where they can find and consume organic waste even when it is buried beneath layers of bedding.

The Journal of Ethology: The International Journal of Behavioural Biology published a 2010 study using Eisenia fetida, red wiggler worms, showing that they use touch to "communicate and influence each other's behavior to travel in the same direction." The study further demonstrated that the worms use touch rather than pheromone trails to influence each other's behavior and to communicate. In your own farm, you can be sure that worms will use their heightened sense of touch to find food and tell their friends about it. This is why you can put organic matter in one corner of your bin and soon find all your worms gathered around, happily devouring it.

Environmental impact of vermicomposting

The environmental impact of vermicomposting cannot be under-estimated. As one of the most cost-effective waste management techniques we have available, vermicomposting helps to maintain an ecological balance. Between our chronic overuse of chemicals on crops and in our landscapes, greenhouse gas emissions generated by our daily lives, and oil spills that damage the environment, we need all the help our little worm friends can give us.

VERMICOMPOSTING CAN REDUCE GREENHOUSE GAS EMISSIONS

Vermicomposting can have a significant impact on reducing greenhouse gas emissions. When organic waste is sent to a landfill, it decomposes anaerobically (without oxygen) and produces methane, a potent greenhouse gas. By diverting organic waste to a worm farm, it is decomposed aerobically (with oxygen) and produces significantly less methane. In this way, vermicomposting can help reduce greenhouse gas emissions and combat climate change.

Our World in Data reported in 2020 that food production accounts for about one-quarter of global greenhouse gas emissions. Some of this loss cannot be avoided. We all have to eat, after all. However, 6% of those emissions are strictly due to food waste, and higher if we account for food that is lost during the production and harvesting phase. Since this part of the food production and consumption cycle is something individuals or farmers have direct control over, we can try to reduce the greenhouse emissions ourselves through the use of vermicompost systems (Ritchie, 2020). Whether we are only vermicomposting our own kitchen scraps or using large-scale vermicomposting farms to eat away fruits and vegetables that are lost during the harvesting process, a concerted effort around the world can make a significant impact.

Additionally, it has also been found that regular, thermophilic composting increases greenhouse gases in the atmosphere. *The Journal of Cleaner Production* published a study in 2016 comparing the reduction in greenhouse gas emissions and nitrogen losses between vermicomposting and regular composting, concluding that vermicomposting is a significantly better choice. In this study, it was shown that not only methane but nitrous oxide was substantially reduced through the practice of vermicomposting. Adding earthworms to compost piles could reduce the amount of greenhouse gases released through the composting process (Nigussie et al., 2016).

Greenhouse gas emissions are a growing concern worldwide, and vermicomposting is one of the easiest to use and most cost-effective tools we have to reverse some of that damage. Less food wasted means less methane released into the atmosphere and less worry for all of us.

Worms represent sustainability and closing the input loop on a garden or homestead. Not only do they provide all the fertility that my garden needs, but they dispose of otherwise wasted scraps so I am able to recycle the entirety of my food waste on site."

Elise Pickett
The Urban Gardener

VERMICOMPOSTING HAS BEEN USED TO CLEAN UP OIL SPILLS

Vermicompost has been used to clean up oil spills and other environmental disasters. When worms are added to soil contaminated with oil, they consume the oil and convert it into biomass, which can then be removed and disposed of safely. This process, known as bioremediation or vermiremediation, is a sustainable and cost-effective way to clean up oil spills and other environmental pollutants.

Although small oil spills happen regularly, statistics report an average of 1.8 large oil spills from tanker incidents each year since 2010. Granting a great reduction in significant oil spills in the last few decades, crude oil contaminants being released into the environment are still a considerable problem (Aizarani, 2023). Some of these incidents take place in the transportation of oil, and some are part of the regular activities conducted by the oil sector, such as drilling or refining. Crude oil is made up of a variety of toxic compounds, varying in toxicity to humans and the environment. These oils pose a health risk to both humans and animals and can result in health problems ranging from mild to life-threatening, such as cancer.

First used in the 1972 Sun Oil pipeline spill in Pennsylvania, bioremediation has since been used in countless oil spills with good success, depending on the type of contaminant being cleaned up. In this process, worms use their own microbes to biodegrade contaminated oil in their guts, cleaning up as much as 99% of contaminants in the soil. In a study in Nigeria using Hyperiodrilus africanus, a species of earthworm native to humid tropical Africa regions, it was shown that if earthworms were added to contaminated soil after the contaminants decreased to levels tolerated by the worms, they would then be able to remove much of the remaining contaminants and leave the soil further enriched by vermicast. Some of the contaminants the study worms were able to reduce were benzene, toluene, ethylbenzene, TPH, and xylene (Ekperusi & Aigbodion, 2015).

The Perfect Temp

Red worms prefer an environment between 55 and 77 degrees Fahrenheit. Temperatures below 40 degrees can be fatal to your worms, so bringing your worm farm indoors during cold weather is vital.

Fun Facts About Worm Biology

- Some scientists believe that worms can breathe underwater for up to two weeks. In soil that is too dry or too wet, your worms are more likely to die from a lack of oxygen than from drowning.

- Instead of eyes, earthworms have light receptors, sensors that sit within the skin and are connected to various nerves inside the worms' bodies. These sensors detect the difference between light and dark but don't provide anything similar to the complexity of the human eye.

- Worms breathe through their skin! This is why keeping the moisture levels in a good balance is vital to the success of your vermicompost farm. Dry and compact soil means that your worms will have a hard time moving and breathing in the soil.

- Worms are hermaphrodites, which means that each worm has both male and female body parts. They still require other worms with which to exchange semen, but each worm will subsequently form its own eggs after mating.

References

Aizarani, J. (2023, February 23). Global average oil spills per decade 2022. Statista. https://www.statista.com/statistics/671539/average-number-of-oil-spills-per-decade/#:~:text=There%20was%20an%20average%20of,spills%20has%20been%20notably%20reduced

Arancon, N. Q., Pant, A., Radovich, T., Hue, N. V., Potter, J. K., & Converse, C. E. (2012). Seed germination and seedling growth of tomato and lettuce as affected by vermicompost water extracts (teas). *HortScience, 47*(12), 1722-1728. https://journals.ashs.org/hortsci/view/journals/hortsci/47/12/article-p1722.xml

Bosch, M., & Hawrysh, J. (2019, May 3). *Carbon: Nitrogen 'sweet spot' (25:1) [INFOGRAPHIC]*. Wiggle Room — Vermicomposting and Worm Castings in Connecticut. https://www.wiggleroom.org/faq.html

California Academy of Sciences. (n.d.). *Earthworm herd*. https://www.calacademy.org/explore-science/earthworm-herd

Domínguez, J. (2018). Earthworms and vermicomposting. In Sajay Ray (Ed.), *Earthworms — The Ecological Engineers of Soil* (pp. 63-77). InTechOpen. doi: 10.5772/intechopen.76088

Ekperusi, O. A., & Aigbodion, F. I. (2015). Bioremediation of petroleum hydrocarbons from crude oil-contaminated soil with the earthworm: Hyperiodrilus Africanus. *3 Biotech, 5*(6), 957-965. https://doi.org/10.1007/s13205-015-0298-1

Environmental Protection Agency. (2023, January 27). *Sustainable Management of Food Basics*. https://www.epa.gov/sustainable-management-food/sustainable-management-food-basics#:~:text=EPA%20estimated%20that%20in%202018,amount%20combusted%20with%20energy%20recovery

Gill, P., & Dommalapati, S. R. (2020). Comparative effect of different organic fertilizers with chemically... *Journal of Pharmacognosy and Phytochemistry, 9*(5). https://www.phytojournal.com/archives/2020/vol9issue5/PartAS/9-5-623-432.pdf

Hendrickson, E. (2021, March 29). *12 cheap and easy worm bedding options*. No Waste Nutrition. https://nowastenutrition.com/worm-bedding/

Mathew, A. A. (2018). *A Study on vermicomposting of banana leaf waste. International Journal of Management, Technology and Engineering*, *8*(X), 1141-1148. https://www.ijamtes.org/gallery/156. ijmte%20oct%20as.pdf

Mlblevins. (2010, January 5). *Digestive system of an earthworm*. Biology Wise. https://biologywise.com/earthworm-digestive-system

Nigussie, A., Kuyper, T. W., Bruun, S., & de Neergaard, A. (2016). Vermicomposting as a technology for reducing nitrogen losses and greenhouse gas emissions from small-scale composting. *ScienceDirect*, *139*, 429-439. https://composting.ces.ncsu.edu/vermicomposting-2/vermicomposting-for-households/

Pathma, J. (2012). Microbial diversity of vermicompost bacteria that exhibit useful agricultural traits and waste management potential. *SpringerPlus*, *1*(26). https://springerplus.springeropen.com/articles/10.1186/2193-1801-1-26

Ritchie, H. (2020, March 18). *Food waste is responsible for 6% of global greenhouse gas emissions*. Our World in Data. https://ourworldindata.org/food-waste-emissions

Sherman, R. (2018). *The worm farmer's handbook: Mid- to large-scale vermicomposting for farms, businesses, municipalities, schools, and institutions*. Chelsea Green Publishing.

Sherman, R. (2022). *Vermicomposting for households*. NC State Extension. https://composting.ces.ncsu.edu/vermicomposting-2/vermicomposting-for-households/

www.ingramcontent.com/pod-product-compliance
Lightning Source LLC
Chambersburg PA
CBHW071008120626
46546CB00003B/995